Scattering and Diffraction by Wedges 2

To my beloved grandchildren Marianna and Roberto
V.G.D.

To my family Stefania, Marina and Alessandro for their patience and support, this book is affectionately dedicated
G.L.

Waves and Scattering Set

coordinated by
Jean-Michel L. Bernard

Volume 2

Scattering and Diffraction by Wedges 2

The Wiener-Hopf Solution – Advanced Applications

Vito G. Daniele
Guido Lombardi

ISTE

WILEY

First published 2020 in Great Britain and the United States by ISTE Ltd and John Wiley & Sons, Inc.

ISTE Ltd
27-37 St George's Road
London SW19 4EU
UK

www.iste.co.uk

John Wiley & Sons, Inc.
111 River Street
Hoboken, NJ 07030
USA

www.wiley.com

Library of Congress Control Number: 2020936444

British Library Cataloguing-in-Publication Data
A CIP record for this book is available from the British Library
ISBN 978-1-78630-664-7

MIX
Paper from responsible sources
FSC® C013604

Contents

Preface ix

Introduction xiii

Chapter 4. Exact Solutions for Electromagnetic Impedance Wedges 1

4.1. Introduction 1
4.2. A list of the impedance wedge problems amenable to exact WH solutions . . 9
4.3. Cases involving classical WH equations. 10
4.3.1. WH formulation of the diffraction by an impedance half-plane 11
4.3.2. Exact solutions of the diffraction by an impedance half-plane 18
4.3.3. Exact solution for the full-plane junction at skew incidence 37
4.3.4. Exact solution of the penetrable half-plane problem (the jump) 39
4.3.5. Exact solution of the right-angled wedge scattering problem. 40
4.4. Exact solutions for impedance wedge problems with the GWHE form of section 3.5 – form #1 51
4.4.1. The WH solution of the Malyuzhinets problem 52
4.4.2. Diffraction at skew incidence ($\alpha_o \neq 0$) by a wedge with a PEC and a PMC face. 58
4.4.3. Diffraction at skew incidence ($\boldsymbol{\alpha_o \neq 0}$) by a wedge with a PEC face and the other face with diagonal $\boldsymbol{Z}^b$ with one null element 60
4.5. Exact solutions for the impedance wedge problems with the GWHEs written in an alternative form – form #2. 62
4.5.1. Exact factorization with diagonal polynomial matrices $P_{a,b}(m)$ 64
4.5.2. Anisotropic symmetric impedance wedges at normal incidence 67
4.5.3. Non-symmetric wedges at normal incidence with commuting P_a and P_b . 68
4.5.4. Non-symmetric wedges at skew incidence 70
4.5.5. Two particular wedge problems amenable to exact solutions. 72

4.6. A general form of the GWHEs to study the arbitrary face impedance wedges – form #3 76
Appendix 4.A. Some important formulas of decomposition for wedge problems . 79

Chapter 5. Fredholm Factorization Solutions of GWHEs for the Electromagnetic Impedance Wedges Surrounded by an Isotropic Medium 87

5.1. Introduction 87
5.2. Generalized Wiener-Hopf equations for the impenetrable wedge scattering problem of an electromagnetic plane wave at skew incidence 88
5.3. Fredholm factorization solution in the $\bar{\eta}$ plane of GWHEs 92
5.4. Fredholm factorization solution in the $\bar{w}$ plane of GWHEs 95
5.5. Approximate solution of FIEs derived from GWHEs 97
5.6. Analytic continuation of approximate solutions of GWHEs 101
5.7. Far-field computation 103
5.8. Criteria for the examples 110
5.9. Example 1: Symmetric isotropic impedance wedge at normal incidence with Ez polarization 111
5.10. Example 2: Non-symmetric isotropic impedance wedge at normal incidence with Hz polarization and surface wave contribution 119
5.11. Example 3: PEC wedge at skew incidence 121
5.12. Example 4: Arbitrary impedance half-plane at skew incidence 124
5.13. Example 5: Arbitrary impedance wedge at skew incidence 126
5.14. Example 6: Arbitrary impedance concave wedge at skew incidence 128
5.15. Discussion 132
Appendix 5.A. Fredholm properties of the integral equation (5.3.1) 132

Chapter 6. Diffraction by Penetrable Wedges 135

6.1. Introduction 135
6.2. GWHEs for the dielectric wedge at normal incidence (Ez-polarization) 140
6.3. Reduction of the GWHEs for the dielectric wedge at Ez-polarization to Fredholm integral equations 142
6.4. Analytic continuation for the solution of the dielectric wedge at Ez-polarization 154
6.5. Some remarks on the Fredholm integral equations (6.3.24), (6.3.26) and numerical solutions 159
6.6. Field evaluation in any point of the space 162
6.7. The dielectric wedge at skew incidence 165
6.8. Criteria for examples of the scattering by a dielectric wedge at normal incidence (Ez-polarization) 176

6.9. Example: the scattering by a dielectric wedge at normal incidence (Ez-polarization) 177
6.10. Discussion. 186
Appendix 6.A. Fredholm factorization applied to (6.3.2)–(6.3.5) 186
Appendix 6.B. Source term $\overline{n}_i(\overline{\eta})$ 188

References. 199

Index. 205

Summary of Volume 1 209

Preface

The theory of wave diffraction by wedges constitutes one of the fundamental problems in mathematical physics. Beyond the obvious applications to engineering and physics, this topic has generated – and continues to generate – new progress in the techniques of applied mathematics (e.g. the Sommerfeld–Malyuzhinets technique and the Kontorovich–Lebedev method), and applied physics (e.g. the geometrical theory of diffraction and the physical theory of diffraction).

The scattering of wedge structures is of interest in the study of electromagnetic response/interaction of more complex, multiscale computational electromagnetic problems, using diffraction coefficients for modeling subdomains where high-frequency methods hold promise.

This book, which is divided into two volumes, presents a general flexible, novel and powerful methodology for solving problems of electromagnetism in regions containing arbitrary wedge objects; this technique can be extended, as suggested, to different physics.

The proposed technique is based on a generalization of the Wiener-Hopf (WH) technique that allows the study of complex canonical scattering problems.

Research carried out during the past two decades has ordered and systematized the procedure, to obtain spectral equations and integral representations for complex problems to avoid redundancy.

The purpose of this book is to present the application of the so-called generalized Wiener-Hopf technique (GWHT) to wedge scattering problems. This method often

---

This book consists of two volumes; any references to Chapters 1, 2 or 3 can be found in the first volume*: Scattering and Diffraction by Wedges 1.*

adopts very sophisticated mathematical methods; however, since the aim of the authors is to favor the acquisition of fundamental concepts in view of possible applications, they prefer to omit rigorous mathematical proofs for theory such as theorems. For all unproven mathematics, we refer to papers and books that are easily available.

Due to the electrical engineering background of the authors, the two volumes are focused on electromagnetic applications. However all of the material is presented in such a way that extension to other physics and applied mathematical fields is straightforward; from applications in all areas of diffraction problems, such as acoustics, elasticity, aerodynamics, hydrodynamics and so on.

Analogies between some resolved problems in electromagnetics and other areas of physics are reported. In particular, for example, the diffraction of a half-infinite crack in an elastic solid medium is described in detail.

Before dealing with the applications to wedge problems, we present some useful remarks about the WH technique. First, the classical and consolidated method of solution of WH equations is founded on a deep knowledge of the properties of complex functions, which is based on the following classical steps: the multiplicative factorization of the kernel, the additive factorization (decomposition) of functions and the application of Liouville's theorem.

Second, in general, no WH problem is simple to study and solve. The formulation can be cumbersome to obtain and very specialized techniques are needed to implement the factorization procedure. Moreover, these methods frequently do not allow small variations of the problem itself. In this context, it is fundamental to adopt spectral domains using spatial Laplace transforms where all the unknown field representations are analytic functions.

Furthermore, even though formal solutions can be obtained in particular cases, a long and difficult elaboration may be required to make them effective from the physical and engineering points of view. For this purpose, tools such as the saddle point method are necessary for the asymptotic evaluation of integrals.

In this book, we aim to solve these problems systematically by helping the reader to implement solutions using the WH method with precise, simple steps and reducing the restrictions on generalization.

In particular, we present one of the most relevant recent advances in the WH technique, i.e. the transformation of WH equations into integral equations for their effective solution via simple discretization.

This method, known as the Fredholm factorization, is an approximate technique that extends the applicability of the WH method to a new variety of problems, consisting of different geometries and materials.

The Fredholm factorization is a semi-analytical solution method that preserves the spectral properties of the problem through its structural singularities and singularities related to the sources. Approximations are present on correction terms that are only related to diffractive properties. Thus, using a suitable analytic continuation, the method allows the reader to look at the physical properties of the problems, as done in analytical closed-form solution methods.

The success of the Fredholm factorization, in solving complex diffraction problems involving isotropic media, has also encouraged the authors to extend the WH technique developed in this book to wedge problems involving anisotropic or, more general, bi-anisotropic media. Moreover, in these cases, the Fredholm factorization is conceptually possible. However, in most cases, it yields complex kernels that involve cumbersome multivalued functions. Much effort is needed to complete the general WH theory for solving wedge problems involving arbitrary linear media. In particular, difficulties increase considerably when it is not possible to take advantage of the presence of Helmholtz equations. In fact, in these cases, it is very difficult and cumbersome to implement the indispensable analytic continuation procedures.

However, to successfully develop the mathematical theory and applications of this book, it is useful to have a technical computing software, particularly in analytical and numerical manipulations of the equations.

Finally, we admit that the authors are strongly influenced by the fundamental work of Felsen and Marcuvitz on radiation and scattering by waves, and therefore the adopted notations follow their studies. In particular, equivalent network representations of equations are provided to systemize the mathematical procedure. The use of different notations in applied physics, applied mathematics and engineering communities should not present a major difficulty in reading this book.

Vito G. DANIELE
Guido LOMBARDI
May 2020

Introduction

Diffraction by objects presenting edges is a fundamental problem in the theory of diffraction. Its importance has been demonstrated by the vast number of papers and books that deal with this topic. These works concern mathematical and physical aspects, with applications to many branches of physics and engineering such as electromagnetism, acoustics, elasticity, aerodynamics and hydrodynamics. To provide a complete theory of the effects of the edges present in a body, we study wedge problems, i.e. canonical problems where the diffracting body is composed of homogeneous wedges. Wedge problems can be classified in different ways. For example, in electromagnetism, we introduce impenetrable and penetrable wedges. Impenetrable wedges arise from the introduction of approximate boundary conditions on the surfaces of wedges. For example, near a metallic face, at microwave frequencies, we can assume that the tangential electrical field vanishes. The introduction of impenetrable wedges simplifies the study of wedge problems considerably since the external problem (i.e. the evaluation of the electromagnetic field outside the wedge) is decoupled from the internal problem (i.e. the evaluation of the electromagnetic field inside the wedge). Considerable advances have been made to obtain more realistic approximate boundary conditions. Unfortunately, the introduction of sophisticated boundary conditions is not sufficient to study penetrable wedges. In particular, the scattering by a penetrable wedge (e.g. the dielectric wedge in electromagnetics) is a very difficult problem that has arisen for a long time and drawn the interest of many scientists.

Nevertheless, even for the simplest cases, no closed-form solution for this problem is known.

This book consists of two volumes; any references to Chapters 1, 2 or 3 can be found in the first volume: *Scattering and Diffraction by Wedges 1.*

The solutions of wedge problems have a long history, which have been well documented by numerous books and papers. This history also documents the importance of the use of spectral representations to obtain the solutions of wedge problems. One of the aims of these two volumes is to claim the power of the representations of the field spectra through Laplace transforms in radial direction. In particular, the use of these representations allows the extension of the powerful WH technique to formulate and efficiently solve many wedge problems.

This book is divided into two volumes that are consistent with each other. The first volume introduces the general theory with simple examples, while the second volume reports more advanced applications. Each volume consists of three chapters and their cross-references.

In each volume, sections, equations and figures are numbered as (chapter.section. item) in order for the content to be easily accessible.

Chapter 1 starts by recalling some fundamental properties of the mathematical theory of the WH technique. The separation of the plus and minus Laplace transforms is fundamental in introducing the concept of the classical factorization, an ingenious idea forming the basis of this technique. This chapter presents a few cases where exact factorizations are available. A novel aspect presented in this chapter is the introduction of the solution procedure for the Fredholm factorization, as an alternative to the classical WH solution procedure. It presents the fundamental advantage of its applicability in general cases where the classical factorizations are not available.

To discuss the importance of the mathematical tools described in Chapter 1, the solution of the most simple diffraction problem is presented in Chapter 2: the electromagnetic scattering of a plane wave by a perfectly electrically conducting (PEC) half-plane. Both the classical factorization procedure and the Fredholm factorization method are applied to obtain the solution of this problem. The comparison of these two different factorization techniques illustrates their advantages and disadvantages. To provide a general methodology for studying basic problems in other diffraction areas, Chapter 2 presents the complete theory of the diffraction of a half-infinite crack located in an indefinite elastic solid.

Chapter 3 deals with the application of the WH technique to wedge problems. This application comes from the introduction of generalized Wiener-Hopf equations (GWHEs) and from the mapping that reduces GWHEs to classical WH equations (CWHEs).

In order to compare the WH technique with the Sommerfeld–Malyuzhinets technique, this chapter introduces two angular complex planes called w and bar w.

In this framework, we present a new tool called rotating waves to study angular regions. Moreover, the introduction of angular complex planes is also useful in developing the application of the Fredholm factorization method.

Chapter 4 (Volume 2) deals with the solution of impenetrable wedges immersed in free space. Exact solutions obtained with the classical factorization theory are compared with those obtained by the Sommerfeld–Malyuzhinets method.

Chapter 5 (Volume 2) is particularly important for the development of immediate engineering applications. In fact, it presents the application of the Fredholm factorization method to the solution of the general scattering problem by an arbitrary impedance wedge illuminated by plane waves at skew incidence.

Finally, Chapter 6 (Volume 2) deals with the use of the WH technique to solve an arbitrary homogeneous penetrable wedge immersed in free space. In particular, this chapter studies, in depth, the diffraction by a dielectric wedge illuminated by plane waves at normal and skew incidence. The WH solution of the dielectric wedge problem presented in this chapter is the first one that the authors of this book have developed in the literature. It contains very sophisticated mathematical methods to overcome difficult analytical problems related to the presence of two media. Recently, many improvements have been made to the WH method to study the diffraction of complex problems consisting of multiple wedges and layers (see 2016–2019 papers by the authors of this book). We expect that these improvements will allow a more simple mathematical procedure to solve the diffraction by a dielectric wedge.

4

Exact Solutions for Electromagnetic Impedance Wedges

4.1. Introduction

While we considered some particular case in Chapter 3, this chapter focuses on the solution of several impedance wedge problems using exact factorization in WH equations. In particular, we will discuss the cases where the considered geometries are also amenable to exact solutions using the Sommerfeld–Malyuzhinets (SM) method. We note that each wedge problem requires some efforts to be solved using both the WH technique and the SM technique. In this context, while comparing the WH solutions proposed in this chapter with the solutions obtainable by the SM technique, we recall that SM solutions are well documented in the literature with great mathematical details. However, due to restricted space, in this chapter, we limit the presentations of the exact WH solutions by delineating the main aspects. We hope that the information presented in this chapter should be sufficient for the interested reader to reproduce the results and to develop all the details that are necessary for practical applications. This is possible because a rigorous mathematical theory of the factorization of arbitrary matrices provides uniqueness and existence theorems.

Figure 4.1.1 presents the scattering problem that is described using Cartesian coordinates along with cylindrical coordinates. We assume the wedge with cylindrical symmetry with respect to the z axis.

This book consists of two volumes; any references to Chapters 1, 2 or 3 can be found in the first volume: *Scattering and Diffraction by Wedges 1*.

To study the characteristic of the WH matrices amenable to the exact classical factorization in the $\bar{\eta}$ plane, as defined in mapping (3.4.1), we consider matrix (4.1.1) obtained in section 4.6 for the general impedance wedge problem, shown in Figure 4.1.1, where $\gamma_b = \gamma_a = \Phi$:

$$\begin{vmatrix} \mathbf{1} & \frac{1}{\bar{\xi}_-}\bar{\mathbf{f}}(\bar{\eta}) \\ \bar{\xi}_-\bar{\mathbf{f}}(\bar{\eta}) & \mathbf{1} \end{vmatrix} \tag{4.1.1}$$

In this matrix of order 4, the quantity $\bar{\mathbf{f}}(\bar{\eta})$ is a matrix of order 2 and the scalar $\bar{\xi}_-(\bar{\eta})$ is defined by

$$\bar{\xi}_-(\bar{\eta}) = \sqrt{\frac{\tau_o + \bar{\eta}}{2}} \tag{4.1.2}$$

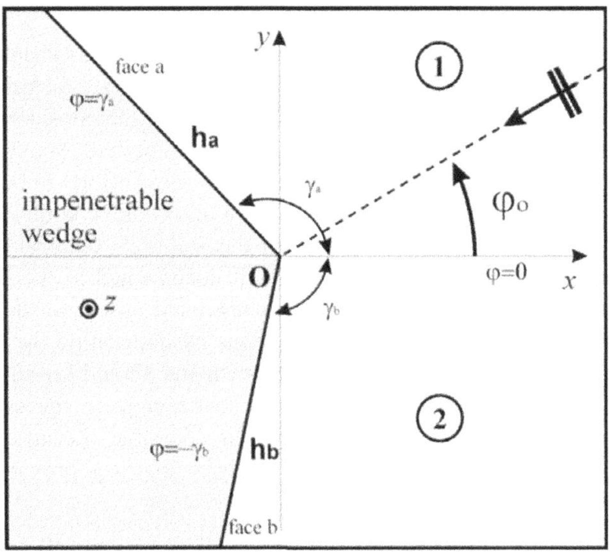

Figure 4.1.1. *Scattering by an impendence wedge*

If the matrix $\bar{\mathbf{f}}(\bar{\eta})$ is diagonal, a commutative algebra is involved in the factorization of (4.1.1). Following section 1.8 and taking into account that

$$\frac{a}{b} = \frac{1}{\bar{\xi}_-^2} = \frac{2}{\tau_o + \bar{\eta}} \tag{4.1.3}$$

is rational in $\bar{\eta}$, equations (1.8.3), (1.8.4) and (1.8.10) can be used to obtain the factorized matrices of (4.1.1). More generally, assuming the Jordan representation of $\bar{\mathbf{f}}(\bar{\eta})$:

$$\bar{\mathbf{f}}(\bar{\eta}) = S(\bar{\eta}) \cdot d_f(\bar{\eta}) \cdot S^{-1}(\bar{\eta}) \tag{4.1.4}$$

with $S(\bar{\eta})$ and $d_f(\bar{\eta})$ being respectively the eigenvector matrix and the eigenvalue matrix of $\bar{\mathbf{f}}(\bar{\eta})$, we can rewrite matrix (4.1.1) in the form:

$$\begin{vmatrix} S^{-1} & 0 \\ 0 & S^{-1} \end{vmatrix} \begin{vmatrix} \mathbf{1} & \frac{1}{\xi_-}\mathbf{f} \\ \xi_-\mathbf{f} & \mathbf{1} \end{vmatrix} \begin{vmatrix} S & 0 \\ 0 & S \end{vmatrix} = \begin{vmatrix} \mathbf{1} & \frac{1}{\xi_-}\mathbf{d}_f \\ \xi_-\mathbf{d}_f & \mathbf{1} \end{vmatrix} \tag{4.1.5}$$

This means that provided that the eigenvector S is constant, the matrix (4.1.1) can again be exactly factorized because the Jordan matrix $\mathbf{d}_f$ of the eigenvalues of $\bar{\mathbf{f}}(\bar{\eta})$ always introduce a commutative algebra. A very simple case where the eigenvectors of $\bar{\mathbf{f}}(\bar{\eta})$ are constant is when $\bar{\mathbf{f}}(\bar{\eta})$ commutes with a constant matrix **C**. It yields:

$$\bar{\mathbf{f}}(\bar{\eta}) = f_o(\bar{\eta})\mathbf{1} + f_1(\bar{\eta})\mathbf{C} \tag{4.1.6}$$

where the scalars $f_o(\bar{\eta})$ and $f_1(\bar{\eta})$ are arbitrary functions and **C** is a constant matrix of order 2.

Finally, all the impedance wedge problems involving diagonal $\bar{\mathbf{f}}(\bar{\eta})$, or more generally, having constant eigenvector matrices, can be solved in closed form using the WH technique. A similar result has been obtained by Bernard (1998) in the framework of the SM method. In particular, impedance wedges that allow the exact solutions of SM difference equations yield a function $\bar{\mathbf{f}}(\bar{\eta})$ that has constant eigenvectors (see section 4.5.4).

Let us now consider the particular cases of impedance wedges shown in Figure 4.1.2. It is remarkable that for these wedge geometries, there is no need to introduce the $\bar{\eta}$-plane defined by (3.4.1). In fact, we always deal with classical WH equations (Daniele 2003b; Daniele and Zich 2014) defined in the η-plane.

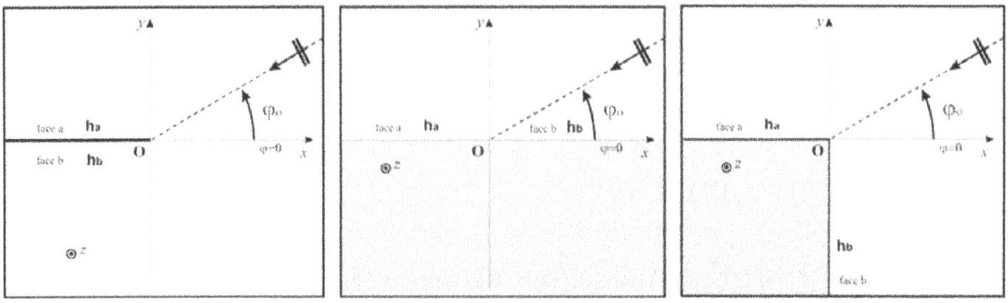

Figure 4.1.2. *Particular cases of impedance wedges. Left: half-plane ($\gamma_a = \gamma_b = \pi$). Center: half-plane junction ($\gamma_a = \pi, \gamma_b = 0$). Right: right wedge ($\gamma_a = \pi, \gamma_b = \pi/2$). Figures 4.1.2 and 4.3.1 describe the same problem but with different definitions of boundary conditions*

Using the general property that, in the factorization processes, the possibility of having exact factorization is not influenced by pre- or post-multiplications by rational matrices (Daniele and Zich 2014), it is remarkable that, for these particular wedges, the general matrices to be factorized have the form

$$\mathbf{G}(\eta) = \mathbf{1} + f(\eta)\mathbf{P}(\eta) \tag{4.1.7}$$

where $f(\eta)$ is a scalar function and $\mathbf{P}(\eta)$ is a polynomial matrix whose degree depends on the problem under examination.

A general theory to factorize the above matrices is presented in Daniele (1984b) and Daniele and Zich (2014). This theory is based on the observation that the logarithmic decomposition of $\mathbf{G}(\eta)$ involves a commutative algebra. Unfortunately, in most cases, the factorized matrices obtained by Daniele (1984a) are not regular (see section 1.5), i.e. not bounded as $\eta \to \infty$. A procedure to render them regular was proposed in Daniele (1984a) and Daniele and Zich (2014). In most cases, this procedure is very cumbersome and always requires the solution of nonlinear algebraic equations. We illustrate this procedure for some cases of diffraction in sections 4.3.2.2 and 4.3.2.6. Henceforth, we will use "not simple" to refer to all the cases that require a regulation procedure to eliminate the unbounded behavior at $\eta \to \infty$. With reference to the SM technique, "the not simple" problems yield difference equations that are of order 2 or more. In particular, exact solution procedures to solve difference equations of order greater than 1 were proposed in Senior and Legault (2000) and Antipov and Silvestrov (2004a, 2004b). Moreover, in these studies, the need to solve nonlinear algebraic equations is evident. While these studies dealt with difference equations that resulted from wedge problems, the importance of formulations in terms of difference equations and their general solution procedure was discussed in Bernard (2014), which analyzed more complex canonical problems: the impedance polygon and the cone.

To recall the WH equations that occur in impenetrable wedge problems, we reconsider the functional equations presented in section 3.3 that are useful to obtain the generalized Wiener-Hopf equations (GWHEs) of wedge problems with the impedance boundary condition (Figure 4.1.1).

A plane wave with skew incidence $\hat{n}_i$ excites an impedance wedge where linear conditions are defined on the boundaries $\varphi = \gamma_a$ and $\varphi = -\gamma_b$. The longitudinal field components of the incident plane wave are given by

$$E_z^i = E_o e^{j\tau_o \rho \cos(\varphi - \varphi_o)} e^{-j\alpha_o z}, \quad H_z^i = H_o e^{j\tau_o \rho \cos(\varphi - \varphi_o)} e^{-j\alpha_o z} \tag{4.1.8}$$

where E_o and H_o are known quantities, ϑ_o is the angle between $\hat{n}_i$ and $\hat{z}$, $\alpha_o = k\cos\vartheta_o$ and $\tau_o = k\sin\vartheta_o$.

The most general linear impedance conditions on the boundary of a material are defined by Senior and Volakis (1995):

$$\begin{bmatrix} E_z \\ H_z \end{bmatrix} = \begin{bmatrix} Z_e(\frac{\partial}{\partial z}, \frac{\partial}{\partial \rho}) & T_e(\frac{\partial}{\partial z}, \frac{\partial}{\partial \rho}) \\ T_h(\frac{\partial}{\partial z}, \frac{\partial}{\partial \rho}) & Y_h(\frac{\partial}{\partial z}, \frac{\partial}{\partial \rho}) \end{bmatrix} \begin{bmatrix} H_\rho \\ -E_\rho \end{bmatrix} \tag{4.1.9}$$

where the elements of the matrix depend on the considered material according to the different levels of approximations derived by Senior and Volakis (1995). The first level of approximation yields the elements Z_e, Y_h, T_e, T_h, which are constant parameters that do not depend on the operators $\partial/\partial z$ and $\partial/\partial\rho$. In this book, this level of approximation is the only one taken into account. The presence of the operators $\partial/\partial z$ and $\partial/\partial\rho$ on the boundaries of the wedges introduces spatial dispersivity, which requires modifications of the theory developed in this chapter.

On the boundaries of the wedge $\varphi = \gamma_a$ and $\varphi = -\gamma_b$, we assume the conditions:

$$\begin{bmatrix} E_z(\rho,\gamma_a) \\ H_z(\rho,\gamma_a) \end{bmatrix} = h_a \begin{bmatrix} H_\rho(\rho,\gamma_a) \\ -E_\rho(\rho,\gamma_a) \end{bmatrix}, \begin{bmatrix} E_z(\rho,-\gamma_b) \\ H_z(\rho,-\gamma_b) \end{bmatrix} = -h_b \begin{bmatrix} H_\rho(\rho,-\gamma_b) \\ -E_\rho(\rho,-\gamma_b) \end{bmatrix} \tag{4.1.10}$$

with the matrices $h_{a,b} = \begin{bmatrix} Z_e^{a,b} & T_e^{a,b} \\ T_h^{a,b} & Y_h^{a,b} \end{bmatrix}$ depending on the wedge material.

The Maxwell equations provide the following relations between the radial components E_ρ and H_ρ and the longitudinal components E_z and H_z:

$$\rho E_\rho = \frac{1}{j\tau_o^2}\left[\alpha_o \rho \frac{\partial E_z}{\partial \rho} + \omega\mu \frac{\partial H_z}{\partial \varphi}\right], \ \rho H_\rho = \frac{1}{j\tau_o^2}\left[\alpha_0 \rho \frac{\partial H_z}{\partial \rho} - \omega\varepsilon \frac{\partial E_z}{\partial \varphi}\right] \quad (4.1.11)$$

We observe that (4.1.10) is an alternative representation of the boundary conditions (4.1.12) on the faces a and b used in section 3.5 of Chapter 3 (3.5.1):

$$\begin{bmatrix} E_z(\rho,\gamma_a) \\ E_\rho(\rho,\gamma_a) \end{bmatrix} = Z^a \begin{bmatrix} H_\rho(\rho,\gamma_a) \\ -H_z(\rho,\gamma_a) \end{bmatrix}, \begin{bmatrix} E_z(\rho,-\gamma_b) \\ E_\rho(\rho,-\gamma_b) \end{bmatrix} = -Z^b \begin{bmatrix} H_\rho(\rho,-\gamma_b) \\ -H_z(\rho,-\gamma_b) \end{bmatrix} \quad (4.1.12)$$

with the matrices $Z^{a,b} = Z_o \begin{bmatrix} z_{11}^{a,b} & z_{12}^{a,b} \\ z_{21}^{a,b} & z_{22}^{a,b} \end{bmatrix}$ depending on the wedge material.

The impedance matrix Z is related to the matrix h and vice versa through the equations:

$$Z_o z_{11} = \frac{Z_e Y_h - T_e T_h}{Y_h}, \ Z_o z_{12} = -\frac{T_e}{Y_h}, \ Z_o z_{21} = \frac{T_h}{Y_h}, \ Z_o z_{22} = \frac{1}{Y_h} \quad (4.1.13)$$

$$Z_e = Z_o \frac{z_{11} z_{22} - z_{12} z_{21}}{z_{22}}, \ T_e = -\frac{z_{12}}{z_{22}}, \ T_h = \frac{z_{21}}{z_{22}}, \ Y_h = \frac{1}{Z_o z_{22}} \quad (4.1.14)$$

The mathematical problem to be solved consists of finding the solutions E_z and H_z of the wave equations $\frac{\partial^2}{\partial x^2} E_z + \frac{\partial^2}{\partial y^2} E_z + \tau_o^2 E_z = 0$, $\frac{\partial^2}{\partial x^2} H_z + \frac{\partial^2}{\partial y^2} H_z + \tau_o^2 H_z = 0$ such that (4.1.10) or (4.1.12) hold on the boundaries $\varphi = \gamma_a$ and $\varphi = -\gamma_b$ of the wedge, as well as $E_z - E_z^i$ and $H_z - H_z^i$ satisfy the Sommerfeld radiation conditions in the angular region $-\gamma_b \le \varphi \le \gamma_a$. In physical problems, the materials of the wedge yield parameters of $h_{a,b}$ that guarantee the existence and uniqueness of the solution.

Equations (3.3.57)–(3.3.60) and the Leontovich conditions (4.1.10), written in the Laplace domain, yield the following four equations:

$$\xi V_{z+}(\eta,0)-\frac{\tau_o^2}{\omega\varepsilon}I_{\rho+}(\eta,0)-\frac{\alpha_o\eta}{\omega\varepsilon}I_{z+}(\eta,0)=(a_1\eta+b_1\,\xi+e_1\,)I_{\rho+}(-m_a,\gamma_a) \\ +(c_1\eta+d_1\,\xi+f_1\,)V_{\rho+}(-m_a,\gamma_a) \tag{4.1.15}$$

$$\xi V_{z+}(\eta,0)+\frac{\tau_o^2}{\omega\varepsilon}I_{\rho+}(\eta,0)+\frac{\alpha_o\eta}{\omega\varepsilon}I_{z+}(\eta,0)=(a_2\,\eta+b_2\,\xi+e_2)I_{\rho+}(-m_b,-\gamma_b) \\ +(c_2\,\eta+d_2\,\xi+f_2)V_{\rho+}(-m_b,-\gamma_b) \tag{4.1.16}$$

$$\xi\, I_{z+}(\eta,0)+\frac{\tau_o^2}{\omega\mu}V_{\rho+}(\eta,0)+\frac{\alpha_o\eta}{\omega\mu}V_{z+}(\eta,0)=(a_3\eta+b_3\,\xi+e_3\,)I_{\rho+}(-m_a,\gamma_a) \\ +(c_3\eta+d_3\,\xi+f_3\,)V_{\rho+}(-m_a,\gamma_a) \tag{4.1.17}$$

$$\xi\, I_{z+}(\eta,0)-\frac{\tau_o^2}{\omega\mu}V_{\rho+}(\eta,0)-\frac{\alpha_o\eta}{\omega\mu}V_{z+}(\eta,0)=(a_4\,\eta+b_4\,\xi+e_4)I_{\rho+}(-m_b,-\gamma_b) \\ +(c_4\,\eta+d_4\,\xi+f_4)V_{\rho+}(-m_b,-\gamma_b) \tag{4.1.18}$$

where $\xi=\xi(\eta)=\sqrt{\tau_o^2-\eta^2}$ $m_{a,b}=m_{a,b}(\eta)=-\eta\cos\gamma_{a,b}+\xi\sin\gamma_{a,b}$ and the constants a_i, b_i, c_i, d_i, e_i, f_i (i=1,2,3,4) depends on the geometrical and electromagnetic parameters and are thus specified by

$$a_1=Z_e^a\sin\gamma_a-\frac{T_h^a\,\alpha_o\cos\gamma_a}{\omega\varepsilon},\ b_1=Z_e^a\cos\gamma_a+\frac{T_h^a\,\alpha_o\sin\gamma_a}{\omega\varepsilon},\ e_1=-\frac{\tau_o^2}{\omega\varepsilon}$$

$$c_1=-T_e^a\sin\gamma_a+\frac{Y_h^a\,\alpha_o\cos\gamma_a}{\omega\varepsilon},\ d_1=-T_e^a\cos\gamma_a-\frac{Y_h^a\,\alpha_o\sin\gamma_a}{\omega\varepsilon}\quad f_1=0$$

$$a_2=-Z_e^b\sin\gamma_b-\frac{T_h^b\,\alpha_o\cos\gamma_b}{\omega\varepsilon},\ b_2=-Z_e^b\cos\gamma_b+\frac{T_h^b\,\alpha_o\sin\gamma_b}{\omega\varepsilon},\ e_2=\frac{\tau_o^2}{\omega\varepsilon}$$

$$c_2=T_e^b\sin\gamma_a+\frac{Y_h^b\,\alpha_o\cos\gamma_b}{\omega\varepsilon},\ d_2=T_e^b\cos\gamma_b-\frac{Y_h^b\,\alpha_o\sin\gamma_b}{\omega\varepsilon}\quad f_2=0$$

$$a_3=T_h^a\sin\gamma_a+\frac{Z_e^a\,\alpha_o\cos\gamma_a}{\omega\mu},\ b_3=T_h^a\cos\gamma_a-\frac{Z_e^a\,\alpha_o\sin\gamma_a}{\omega\mu},\ e_3=0$$

$$c_3 = -Y_h^a \sin\gamma_a - \frac{T_e^a \alpha_o \cos\gamma_a}{\omega\mu},\ d_3 = -Y_h^a \cos\gamma_a + \frac{T_e^a \alpha_o \sin\gamma_a}{\omega\mu}\ \ f_3 = \frac{\tau_o^2}{\omega\mu}$$

$$a_4 = -T_h^b \sin\gamma_b + \frac{Z_e^b \alpha_o \cos\gamma_b}{\omega\mu},\ b_4 = -T_h^b \cos\gamma_b - \frac{Z_e^b \alpha_o \sin\gamma_b}{\omega\mu},\ e_4 = 0$$

$$c_4 = Y_h^b \sin\gamma_a - \frac{T_e^b \alpha_o \cos\gamma_b}{\omega\mu},\ d_4 = Y_h^b \cos\gamma_b + \frac{T_e^b \alpha_o \sin\gamma_b}{\omega\mu}\ \ f_4 = -\frac{\tau_o^2}{\omega\mu} \tag{4.1.19}$$

Note that in addition to the four functions $V_{z+}(\eta,0)$, $I_{z+}(\eta,0)$, $V_{\rho+}(\eta,0)$, $I_{\rho+}(\eta,0)$ that are the plus function in the η plane, there are four other functions which are regular in m_a-lower half-planes and m_b-lower half-planes. Equations (4.1.15)–(4.1.18) can be rewritten in the matrix form:

$$S(\eta)\begin{vmatrix} V_{z+}(\eta,0) \\ I_{\rho+}(\eta,0) \\ I_{z+}(\eta,0) \\ V_{\rho+}(\eta,0) \end{vmatrix} = D(\eta)\begin{vmatrix} I_{\rho+}(-m_a,\gamma_a) \\ I_{\rho+}(-m_b,\gamma_b) \\ V_{\rho+}(-m_a,\gamma_a) \\ V_{\rho+}(-m_b,\gamma_b) \end{vmatrix} \tag{4.1.20}$$

where the identifications of the entries of the matrices $S(\eta)$ and $D(\eta)$ are clear. It yields the GWHEs

$$G(\eta)\begin{vmatrix} V_{z+}(\eta,0) \\ I_{\rho+}(\eta,0) \\ I_{z+}(\eta,0) \\ V_{\rho+}(\eta,0) \end{vmatrix} = \begin{vmatrix} I_{\rho+}(-m_a,\gamma_a) \\ I_{\rho+}(-m_b,\gamma_b) \\ V_{\rho+}(-m_a,\gamma_a) \\ V_{\rho+}(-m_b,\gamma_b) \end{vmatrix} \tag{4.1.21}$$

with $G(\eta) = D^{-1}(\eta)S(\eta)$. Assuming that $\gamma_a = \gamma_b = \Phi$, $S(\eta)$ and $D(\eta)$, $G(\eta) = D^{-1}(\eta)S(\eta)$ are given in section 3.5.

The GWHEs considered here define a closed mathematical problem. Their solutions yield the tangential fields in the directions $\varphi = 0$, $\varphi = \gamma_a$ and $\varphi = \gamma_b$.

The evaluation of the Laplace transforms of the electromagnetic field for each direction φ is a complex problem for the presence of multivalued functions of η. Conversely, working in the w plane (see section 1.13 of Chapter 1), the representations that evaluate the spectra for each direction φ in terms of the spectra in the direction $\varphi = 0$ are simple. In particular, for these representations, we can use the equations given in section 3.10 of Chapter 3.

4.2. A list of the impedance wedge problems amenable to exact WH solutions

To help the reader understand the impedance wedge problems that can be solved exactly by the WH technique, we present a list in this section. The order of the list follows the order of the sections where the solutions are presented. In general, the kernel matrices to be factorized are of order 4. The parameters of these matrices involve matrices of order 2, defined by the impedances Z_a and Z_b and the auxiliary quantities P_a and P_b defined on the two faces a and b, as shown in Figure 4.1.1. By "not simple" problem, we mean that the exact factorization requires the solution of nonlinear algebraic equations. This list is not exhaustive; for example, some canonical problems with rectangular geometry that were solved in Daniele and Zich (2014) are not listed here. In the problems we assume plane wave illumination (4.1.8) at skew ($\alpha_o \neq 0$) or normal ($\alpha_o = 0$) incidence and we classify them in terms of the impedances $\mathbf{Z}_a$ and $\mathbf{Z}_b$ (see section 4.3.1.1 for definitions) and of the polynomial matrices P_a and P_b (see section 4.5.1 for definitions):

a) skew incidence on a half-plane with isotropically different $\mathbf{Z}_a$ and $\mathbf{Z}_b$ (section 4.3.2.1);

b) skew incidence on a half-plane with symmetric $\mathbf{Z}_a = \mathbf{Z}_b$ ("not simple" problem) (section 4.3.2.2);

c) skew incidence on a half-plane with asymmetric $\mathbf{Z}_a = -\mathbf{Z}_b$ (section 4.3.2.3);

d) skew incidence on a half-plane with particular non-invertible $\mathbf{Z}_a$ and $\mathbf{Z}_b$ (section 4.3.2.4);

e) normal incidence on a half-plane with diagonal impedances $\mathbf{Z}_a$ and $\mathbf{Z}_b$ (section 4.3.2.5);

f) skew incidence on a half-plane with scalar reflection coefficients of the two impedances $\mathbf{Z}_a$ and $\mathbf{Z}_b$ ("not simple" problem) (section 4.3.2.6);

g) skew incidence on a half-plane with $\mathbf{Z}_a$=0 and $\mathbf{Z}_b$ diagonal with one vanishing element (section 4.3.2.7)

h) skew incidence on a full-plane junction with the scalar ratio of impedances $\mathbf{Z}_a$ and $\mathbf{Z}_b$ (section 4.3.3);

i) skew incidence on a full-plane junction with $\mathbf{Z}_a$=0 and non-invertible $\mathbf{Z}_b$ (section 4.3.3);

j) skew incidence on a penetrable half-plane (section 4.3.4);

k) skew incidence on a right wedge with $\mathbf{Z}_a$=0 and diagonal $\mathbf{Z}_b$ (section 4.3.5);

l) normal incidence on a wedge with isotropic $\mathbf{Z}_a$ and $\mathbf{Z}_b$, i.e. the Malyuzhinets problem (section 4.4.1);

m) skew incidence on a wedge with $\mathbf{Z}_a$=0 and $\mathbf{Y}_b$=0 (section 4.4.2);

n) skew incidence on a wedge with $\mathbf{Z}_a$=0 and $\mathbf{Z}_b$ diagonal with one vanishing element (section 4.4.3);

o) normal incidence on a wedge with diagonal P_a and P_b (section 4.5.1);

p) normal incidence on a wedge with $P_a = P_b$ (section 4.5.2);

q) normal incidence on a wedge with commuting P_a and P_b (section 4.5.3);

r) skew incidence on a wedge with commuting P_a and P_b and constant eigenvectors (section 4.5.4);

s) skew incidence on a wedge with particular symmetries between the two impedance faces (section 4.5.5).

We observe that the introduction of new constructive exact factorizations of WH matrices is important in itself. In fact, in general, exact analytical solutions provide validation of alternative methods such as the numerical ones. However, when a quick and efficient solution of practical problems is needed, we suggest the application of the Fredholm factorization method, see Chapter 5 for wedge problems. As discussed in Chapter 1, the Fredholm factorization reduces the classical factorization to the solution of a Fredholm integral equation of the second kind. The authors recall that it is also possible to link the solutions of difference equations in the SM method to Fredholm integral equations, as reported in (Rubinowicz 1966; Budaev 1995; Bernard 1997; Lyalinov and Zhu 2003a). In particular, the integral equation approach for the diffraction by a wedge with impedance faces was used in Lyalinov and Zhu (2013).

4.3. Cases involving classical WH equations

As mentioned previously, in some important cases, (4.1.15)–(4.1.18) considerably simplify and yield classical WH equations without resorting to the mapping defined in (3.4.1). These cases are shown in Figures 4.1.2 and 4.3.1, which will be discussed in some detail. While in this book the classical WH equations of the geometries of Figures 4.1.2 and 4.3.1 are first deduced from the angular functional equations (3.3.57)–(3.3.60), we recall that an alternative formulation is possible (Daniele and Zich 2014) based on circuit models.

We will show that the WH equations of the geometry presented in Figure 4.3.1 always involve WH kernels that commute with polynomial matrices. Thus, in principle, matrix factorizations can be achieved in closed form (see Chapter 4 of Daniele and Zich 2014). However, we must note that, in most cases, the relevant factorization formulas are not regular due to the introduction of unbounded behaviors for $\eta \to \infty$. A cumbersome procedure that requires the solution of nonlinear algebraic equations is necessary to make the factorized matrices bounded. Fortunately, for several cases, this procedure is not necessary. It is important to observe that, in these more simple cases, the SM method, when applicable, introduces difference equations of first order amenable to exact solutions.

4.3.1. *WH formulation of the diffraction by an impedance half-plane*

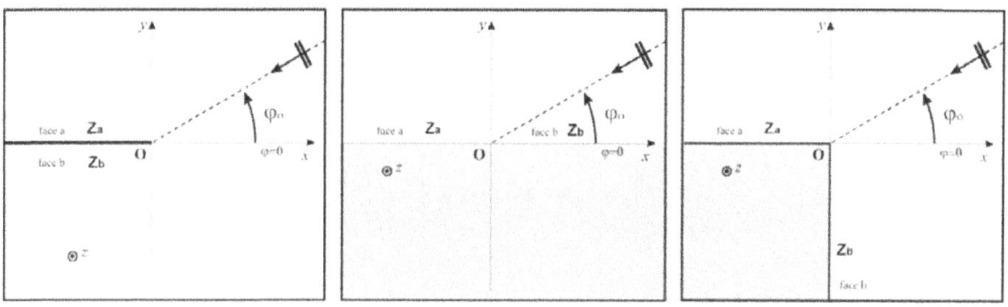

Figure 4.3.1. *Particular cases of impedance wedges. Left: half-plane ($\gamma_a = \gamma_b = \pi$). Center: half-plane junction ($\gamma_a = \pi, \gamma_b = 0$). Right: right wedge ($\gamma_a = \pi, \gamma_b = \pi/2$). Figures 4.1.2 and 4.3.1 describe the same problem but with different definitions of boundary conditions*

Let us consider an incident plane wave that impinges on an impedance half-plane (Figure 4.3.1, left) at skew incidence with the zenithal angle β :

$$E_z^i = E_o e^{j\tau_o \rho\cos(\varphi-\varphi_o)} e^{-j\alpha_o z} \quad H_z^i = H_o e^{j\tau_o \rho\cos(\varphi-\varphi_o)} e^{-j\alpha_o z} \tag{4.3.1.1}$$

where k is the propagation constant and $\alpha_o = k\sin\beta$, $\tau_o = k\cos\beta = \sqrt{k^2-\alpha_o^2}$ $\rho\cos(\varphi-\varphi_o) = x\cos\varphi_o + y\sin\varphi_o$. Using the Maxwell equations, we obtain the following components of the incident plane wave:

$$E_x^i = e^{j\tau_o \rho\cos(\varphi-\varphi_o)} e^{-j\alpha_o z} \frac{E_o\alpha_o \cos\varphi_o + Z_o H_o k \sin\varphi_o}{\tau_o} \tag{4.3.1.2}$$

$$E_y^i = e^{j\tau_o \rho\cos(\varphi-\varphi_o)} e^{-j\alpha_o z} \frac{E_o\alpha_o \sin\varphi_o - Z_o H_o k \cos\varphi_o}{\tau_o} \tag{4.3.1.3}$$

$$H_x^i = e^{j\tau_o \rho\cos(\varphi-\varphi_o)} e^{-j\alpha_o z} \frac{-E_o k Y_o \sin\varphi_o + \alpha_o H_o \cos\varphi_o}{\tau_o} \tag{4.3.1.4}$$

$$H_y^i = e^{j\tau_o \rho\cos(\varphi-\varphi_o)} e^{-j\alpha_o z} \frac{E_o k Y_o \cos\varphi_o + H_o \alpha_o \sin\varphi_o}{\tau_o} \tag{4.3.1.5}$$

where Z_o and $Y_o = Z_o^{-1}$ are the impedance and the admittance of the free space.

With reference to Figure 4.3.1, we introduce the following plus and minus functions defined by the Laplace transform of electromagnetic components located in the x,z plane (where the factor $e^{-j\alpha_o z}$ is omitted):

$$\mathbf{V}_+(\eta) = \int_0^\infty \begin{vmatrix} -E_x(x,0) \\ E_z(x,0) \end{vmatrix} e^{j\eta x} dx \quad \mathbf{I}_+(\eta) = \int_0^\infty \begin{vmatrix} H_z(x,0) \\ H_x(x,0) \end{vmatrix} e^{j\eta x} dx \tag{4.3.1.6}$$

$$\mathbf{V}_{a-}(\eta) = \int_{-\infty}^0 \begin{vmatrix} -E_x(x,0_+) \\ E_z(x,0_+) \end{vmatrix} e^{j\eta x} dx \quad \mathbf{I}_{a-}(\eta) = \int_{-\infty}^0 \begin{vmatrix} H_z(x,0_+) \\ H_x(x,0_+) \end{vmatrix} e^{j\eta x} dx \tag{4.3.1.7}$$

$$\mathbf{V}_{b-}(\eta) = \int_{-\infty}^0 \begin{vmatrix} -E_x(x,0_-) \\ E_z(x,0_-) \end{vmatrix} e^{j\eta x} dx \quad \mathbf{I}_{a-}(\eta) = \int_{-\infty}^0 \begin{vmatrix} H_z(x,0_-) \\ H_x(x,0_-) \end{vmatrix} e^{j\eta x} dx \tag{4.3.1.8}$$

Concerning the boundary conditions on the faces a and b, we introduce the impedances $\mathbf{Z}_a$ and $\mathbf{Z}_b$ defined by (4.3.1.9). We observe that these impedances differ from the impedance $Z^{a,b}$ defined by (4.1.12) since there is a different order of the component spectra in $\mathbf{V}_a$ and $\mathbf{V}_b$:

$$\mathbf{V}_{a-}(\eta) = -\mathbf{Z}_a \mathbf{I}_{a-}(\eta), \quad \mathbf{V}_{b-}(\eta) = \mathbf{Z}_b \mathbf{I}_{b-}(\eta) \tag{4.3.1.9}$$

that yield $\mathbf{Z}_{a21} = Z^a{}_{12}, \mathbf{Z}_{a22} = Z^a{}_{11}, \mathbf{Z}_{a12} = Z^a{}_{21}, \mathbf{Z}_{a11} = Z^a{}_{22}$.

Alternatively, it is also convenient to introduce plus and minus functions defined by Daniele and Zich (2014):

$$\mathbf{E}_{a+}(\eta) = \mathbf{V}_+(\eta) + \mathbf{Z}_a \mathbf{I}_+(\eta) \tag{4.3.1.10}$$

$$\mathbf{E}_{b+}(\eta) = \mathbf{V}_+(\eta) - \mathbf{Z}_b \mathbf{I}_+(\eta) \tag{4.3.1.11}$$

$$\mathbf{X}_-(\eta) = \mathbf{V}_{a-}(\eta) - \mathbf{V}_{b-}(\eta) \tag{4.3.1.12}$$

$$\mathbf{Y}_-(\eta) = \mathbf{I}_{a-}(\eta) - \mathbf{I}_{b-}(\eta) \tag{4.3.1.13}$$

Important quantities are the matrices of order 2 related to the spectral impedance and admittance of the free space:

$$\mathbf{Z}_c(\eta) = \frac{Z_o}{k\xi}\begin{vmatrix} k^2-\eta^2 & \eta\,\alpha_o \\ \eta\,\alpha_o & \tau_o^2 \end{vmatrix},\ \mathbf{Y}_c(\eta) = (\mathbf{Z}_c(\eta))^{-1} = \frac{Y_o}{k\xi}\begin{vmatrix} \tau_o^2 & -\eta\,\alpha_o \\ -\eta\,\alpha_o & k^2-\eta^2 \end{vmatrix} \tag{4.3.1.14}$$

where $\xi = \sqrt{\tau_o^2 - \eta^2}$ and, k, Z_0, Y_o are respectively the wavenumber, the impedance and the admittance of the free space.

Depending on the minus and plus functions that are assumed on the faces a and b of the half-plane and on the aperture y=0, x>0, several classical WH formulations for the diffraction by an impedance half-plane were presented by Daniele and Zich (2014), who, for example, report the two alternative formulations (4.3.1.15) and (4.3.1.18) deduced by circuit considerations:

$$\begin{vmatrix} \mathbf{V}_+ \\ \mathbf{I}_+ \end{vmatrix} = \mathbf{W}\begin{vmatrix} \dfrac{\mathbf{Z}_a\mathbf{I}_{a-} - \mathbf{Z}_b\mathbf{I}_{b-}}{2} \\ -\dfrac{\mathbf{I}_{a-} + \mathbf{I}_{b-}}{2} \end{vmatrix} \tag{4.3.1.15}$$

where the matrix to be factorized is

$$\mathbf{W}(\eta) = \mathbf{1} + \begin{vmatrix} \mathbf{Z}_c(\eta) & \mathbf{0} \\ \mathbf{0} & \mathbf{Y}_c(\eta) \end{vmatrix}\mathbf{F} \tag{4.3.1.16}$$

with **F** being a constant matrix of order 4 defined by

$$\mathbf{F} = \begin{vmatrix} [\mathbf{1} + (\mathbf{1} - \mathbf{Y}_a\mathbf{Z}_b)(\mathbf{1} + \mathbf{Y}_a\mathbf{Z}_b)^{-1}]\mathbf{Y}_a & (\mathbf{1} - \mathbf{Y}_a\mathbf{Z}_b)(\mathbf{1} + \mathbf{Y}_a\mathbf{Z}_b)^{-1} \\ -(\mathbf{1} + \mathbf{Z}_a\mathbf{Y}_b)^{-1} & \mathbf{2}(\mathbf{Y}_a + \mathbf{Y}_b)^{-1} \end{vmatrix} \tag{4.3.1.17}$$

and $\begin{vmatrix} \dfrac{\mathbf{Z_a I_{a-}} - \mathbf{Z_b I_{b-}}}{2} \\ -\dfrac{\mathbf{I_{a-}} + \mathbf{I_{b-}}}{2} \end{vmatrix}$ is a minus vector of order 4;

$$\mathbf{G_i}(\eta)\begin{vmatrix} \mathbf{X}_-(\eta) \\ \mathbf{Y}_-(\eta) \end{vmatrix} = \begin{vmatrix} \mathbf{E_{a+}}(\eta) \\ \mathbf{E_{b+}}(\eta) \end{vmatrix} \qquad (4.3.1.18)$$

where:

$$\mathbf{G_i}(\eta) = \begin{vmatrix} \dfrac{1}{2}\{\mathbf{1} + \mathbf{Z_a}[\mathbf{Z_c}(\eta)]^{-1}\} & \dfrac{1}{2}\{\mathbf{Z_a} + \mathbf{Z_c}(\eta)\} \\ -\dfrac{1}{2}\{\mathbf{1} + \mathbf{Z_b}[\mathbf{Z_c}(\eta)]^{-1}\} & \dfrac{1}{2}\{\mathbf{Z_b} + \mathbf{Z_c}(\eta)\} \end{vmatrix} \qquad (4.3.1.19)$$

with **1** being the identity matrix of order 2.

Due to the ordering of the plus and minus functions with respect to the kernel in (4.3.1.15) and (4.3.1.18), the factorizations have the form:

$$\mathbf{W}(\eta) = \mathbf{W}_+(\eta)\mathbf{W}_-(\eta) \qquad (4.3.1.20)$$

$$\mathbf{G}_i(\eta) = \mathbf{G}_{i+}(\eta)\mathbf{G}_{i-}(\eta) \qquad (4.3.1.21)$$

Premultiplying $\mathbf{G_i}(\eta)$ by $\begin{vmatrix} \mathbf{1} & \mathbf{-1} \\ \mathbf{1} & \mathbf{1} \end{vmatrix}$ we reduce the factorization to that of

$$\mathbf{G}_e(\eta) = \begin{vmatrix} \dfrac{1}{2}(\mathbf{Z_a} + \mathbf{Z_b} + 2\mathbf{Z_c}(\eta)) & \dfrac{1}{2}\xi_+\{\mathbf{Z_a} - \mathbf{Z_b}\} \\ \dfrac{1}{2\xi_+}\{\mathbf{Z_a} - \mathbf{Z_b}\} & \dfrac{1}{2}(\mathbf{Z_a} + \mathbf{Z_b} + 2\mathbf{Z_c}(\eta)) \end{vmatrix}$$
$$= \begin{vmatrix} 0 & \xi_+ \\ \dfrac{1}{\xi_+} & 0 \end{vmatrix} \begin{vmatrix} \dfrac{1}{2}\{\mathbf{Z_a} - \mathbf{Z_b}\} & \dfrac{1}{2}\xi_+(\mathbf{Z_a} + \mathbf{Z_b} + 2\mathbf{Z_c}(\eta)) \\ \dfrac{1}{2\xi_+}(\mathbf{Z_a} + \mathbf{Z_b} + 2\mathbf{Z_c}(\eta)) & \dfrac{1}{2}\{\mathbf{Z_a} - \mathbf{Z_b}\} \end{vmatrix} = \mathbf{G}_{e+}(\eta)\mathbf{G}_{e-}(\eta)$$

(4.3.1.22)

where $\xi_+ = \sqrt{\tau_o - \eta}$.

Finally, another alternative form is obtained by algebraic manipulation on (4.3.1.22).

In particular, the matrix $\begin{vmatrix} \frac{1}{2}\{\mathbf{Z}_a - \mathbf{Z}_b\} & \xi_+ \frac{1}{2}(\mathbf{Z}_a + \mathbf{Z}_b + 2\mathbf{Z}_c(\eta)) \\ \frac{1}{\xi_+}\frac{1}{2}(\mathbf{Z}_a + \mathbf{Z}_b + 2\mathbf{Z}_c(\eta)) & \frac{1}{2}\{\mathbf{Z}_a - \mathbf{Z}_b\} \end{vmatrix}$

reduces to the factorization of a matrix similar to (4.1.1):

$$\mathbf{M}(\eta) = \begin{vmatrix} \mathbf{1} & \xi_+ \mathbf{f}(\eta) \\ \frac{1}{\xi_+}\mathbf{f}(\eta) & \mathbf{1} \end{vmatrix} = \mathbf{M}_+(\eta)\mathbf{M}_-(\eta) \tag{4.3.1.23}$$

where:

$$\mathbf{f}(\eta) = \mathbf{\Gamma}_i + \mathbf{Y}_\Delta \mathbf{Z}_c(\eta) \tag{4.3.1.24}$$

and the constant matrices are defined by:

$$Z_\Sigma = \frac{Z_a + Z_b}{2},\ \mathbf{Z}_\Delta = \frac{\mathbf{Z}_a - \mathbf{Z}_b}{\mathbf{2}} = \mathbf{Y}_\Delta^{-1},\ \Gamma = (\mathbf{Z}_a - \mathbf{Z}_b)^{-1}(\mathbf{Z}_a + \mathbf{Z}_b) \tag{4.3.1.25}$$

Each of the above representations can be useful to obtain exact factorizations.

For example, (4.3.1.16) shows that, for an impedance half-plane, we deal with matrices that commute with the matrix

$$\xi \begin{vmatrix} \mathbf{Z}_c(\eta) & \mathbf{0} \\ \mathbf{0} & \mathbf{Y}_c(\eta) \end{vmatrix} \mathbf{F} \tag{4.3.1.26}$$

which is a polynomial matrix.

The above formulations were obtained by Daniele and Zich (2014) from circuit considerations. In this section, we again obtain equation (4.3.1.18) from the angular functional equations (3.3.57)–(3.3.60) in the case of $\gamma_a = \gamma_b = \pi$, where we have $m_a = m_b = \eta$, $n_a = n_b = \xi$.

Equations (3.3.57)–(3.3.60) can be rewritten as

$$\xi \; V_{az}(\eta) - \frac{\tau_o^2}{\omega\varepsilon} I_{a\rho}(\eta) - \frac{\alpha_o \eta}{\omega\varepsilon} I_{az+}(\eta) = 0 \tag{4.3.1.27}$$

$$\xi \, I_{az}(\eta) + \frac{\tau_o^2}{\omega\mu} V_{a\rho}(\eta) + \frac{\alpha_o \eta}{\omega\mu} V_{az}(\eta) = 0 \tag{4.3.1.28}$$

$$\xi \; V_{bz}(\eta) + \frac{\tau_o^2}{\omega\varepsilon} I_{b\rho}(\eta) + \frac{\alpha_o \eta}{\omega\varepsilon} I_{bz}(\eta) = 0 \tag{4.3.1.29}$$

$$\xi \, I_{bz}(\eta) - \frac{\tau_o^2}{\omega\mu} V_{b\rho+}(\eta) - \frac{\alpha_o \eta}{\omega\mu} V_{bz+}(\eta) = 0 \tag{4.3.1.30}$$

where:

$$V_{az}(\eta) = V_{z+}(\eta,0) + V_{z+}(-\eta,\pi), \qquad V_{a\rho}(\eta) = V_{\rho+}(\eta,0) - V_{\rho+}(-\eta,\pi) \tag{4.3.1.31}$$

$$I_{az}(\eta) = I_{z+}(\eta,0) + I_{z+}(-\eta,\pi), \qquad I_{a\rho}(\eta) = I_{\rho+}(\eta,0) - I_{\rho+}(-\eta,\pi) \tag{4.3.1.32}$$

$$V_{bz}(\eta) = V_{z+}(\eta,0) + V_{z+}(-\eta,-\pi), \qquad V_{b\rho}(\eta) = V_{\rho+}(\eta,0) - V_{\rho+}(-\eta,-\pi) \tag{4.3.1.33}$$

$$I_{bz}(\eta) = I_{z+}(\eta,0) + I_{z+}(-\eta,-\pi), \qquad I_{b\rho}(\eta) = I_{\rho+}(\eta,0) - I_{\rho+}(-\eta,-\pi) \tag{4.3.1.34}$$

The above equations can be rewritten in the compact form

$$\begin{vmatrix} -V_{a\rho}(\eta) \\ V_{az}(\eta) \end{vmatrix} = \mathrm{Z_c}(\eta) \cdot \begin{vmatrix} I_{az}(\eta) \\ I_{a\rho}(\eta) \end{vmatrix}, \qquad \begin{vmatrix} -V_{b\rho}(\eta) \\ V_{bz}(\eta) \end{vmatrix} = -\mathrm{Z_c}(\eta) \cdot \begin{vmatrix} I_{bz}(\eta) \\ I_{b\rho}(\eta) \end{vmatrix} \tag{4.3.1.35}$$

Taking into account the boundary conditions at $\varphi = \Phi = \pi$ and $\varphi = -\Phi = -\pi$, we have

$$\begin{vmatrix} V_{\rho+}(-\eta,\pi) \\ -V_{z+}(-\eta,\pi) \end{vmatrix} + \mathrm{Z_a} \begin{vmatrix} I_{z+}(-\eta,\pi) \\ I_{\rho+}(-\eta,\pi) \end{vmatrix} = 0, \quad \begin{vmatrix} V_{\rho+}(-\eta,-\pi) \\ -V_{z+}(-\eta,-\pi) \end{vmatrix} - \mathrm{Z_b} \begin{vmatrix} I_{z+}(-\eta,-\pi) \\ I_{\rho+}(-\eta,-\pi) \end{vmatrix} = 0 \tag{4.3.1.36}$$

where $\mathrm{Z_a}$ and $\mathrm{Z_b}$ are the matrix impedances on the two faces and thus

$$\begin{vmatrix} -V_{a\rho}(\eta) \\ V_{az}(\eta) \end{vmatrix} + \mathrm{Z_a} \begin{vmatrix} I_{az}(\eta) \\ I_{a\rho}(\eta) \end{vmatrix} = \begin{vmatrix} -V_{\rho+}(\eta,0) \\ V_{z+}(\eta,0) \end{vmatrix} + \mathrm{Z_a} \begin{vmatrix} I_{z+}(\eta,0) \\ I_{\rho+}(\eta,0) \end{vmatrix} = \mathrm{E}_{a+}(\eta) \tag{4.3.1.37}$$

$$\begin{vmatrix} -V_{b\rho}(\eta) \\ V_{bz}(\eta) \end{vmatrix} - Z_b \begin{vmatrix} I_{az}(\eta) \\ I_{a\rho}(\eta) \end{vmatrix} = \begin{vmatrix} -V_{\rho+}(\eta,0) \\ V_{z+}(\eta,0) \end{vmatrix} - Z_b \begin{vmatrix} I_{z+}(\eta,0) \\ I_{\rho+}(\eta,0) \end{vmatrix} = E_{b+}(\eta) \tag{4.3.1.38}$$

Equations (4.3.1.35) yield

$$\begin{vmatrix} -V_{a\rho}(\eta) \\ V_{az}(\eta) \end{vmatrix} = Z_c(\eta)[Z_c(\eta)+Z_a]^{-1} E_{a+}(\eta),$$
$$\begin{vmatrix} -V_{b\rho}(\eta) \\ V_{bz}(\eta) \end{vmatrix} = Z_c(\eta)[Z_c(\eta)+Z_b]^{-1} E_{b+}(\eta) \tag{4.3.1.39}$$

$$\begin{vmatrix} I_{az}(\eta) \\ I_{a\rho}(\eta) \end{vmatrix} = [Z_c(\eta)+Z_a]^{-1} E_{a+}(\eta) \quad \begin{vmatrix} I_{bz}(\eta) \\ I_{ba\rho}(\eta) \end{vmatrix} = -[Z_c(\eta)+Z_b]^{-1} E_{b+}(\eta) \tag{4.3.1.40}$$

where $\mathbf{E}_{a+}(\eta)$ and $\mathbf{E}_{b+}(\eta)$ are defined by (4.3.1.10) and (4.3.1.11) and taking into account (4.3.1.12) and (4.3.1.13), these equations are equivalent to:

$$X_-(\eta) = \mathbf{Z_c}(\eta)\left\{[\mathbf{Z_c}(\eta)+\mathbf{Z_a}]^{-1}\mathbf{E}_{a+}(\eta) - [\mathbf{Z_c}(\eta)+\mathbf{Z_b}]^{-1}\mathbf{E}_{b+}(\eta)\right\} \tag{4.3.1.41}$$

$$Y_-(\eta) = [\mathbf{Z_c}(\eta)+\mathbf{Z_a}]^{-1}\mathbf{E}_{a+}(\eta) + [\mathbf{Z_c}(\eta)+\mathbf{Z_b}]^{-1}\mathbf{E}_{b+}(\eta) \tag{4.3.1.42}$$

Due to the presence of the relationship:

$$Z_c(\eta)[Z_c(\eta)+Z_a]^{-1} = (1+Z_a Z_C^{-1}(\eta))^{-1} \tag{4.3.1.43}$$

$$Z_c(\eta)[Z_c(\eta)+Z_b]^{-1} = (1+Z_b Z_C^{-1}(\eta))^{-1} \tag{4.3.1.44}$$

The following representation also holds in the presence of anisotropic Z_a and Z_b:

$$G(\eta) = \begin{vmatrix} \mathbf{Z}_c(\eta)[\mathbf{Z}_c(\eta)+\mathbf{Z}_a]^{-1} & -\mathbf{Z}_c(\eta)[\mathbf{Z}_c(\eta)+\mathbf{Z}_b]^{-1} \\ [\mathbf{Z}_c(\eta)+\mathbf{Z}_a]^{-1} & [\mathbf{Z}_c(\eta)+\mathbf{Z}_b]^{-1} \end{vmatrix}^{-1}$$
$$= \begin{vmatrix} \frac{1}{2}\left\{\mathbf{1}+\mathbf{Z}_a[\mathbf{Z}_c(\eta)]^{-1}\right\} & \frac{1}{2}\{\mathbf{Z_a}+\mathbf{Z}_c(\eta)\} \\ -\frac{1}{2}\left\{\mathbf{1}+\mathbf{Z}_b[\mathbf{Z}_c(\eta)]^{-1}\right\} & \frac{1}{2}\{\mathbf{Z}_b+\mathbf{Z_c}(\eta)\} \end{vmatrix} \tag{4.3.1.45}$$

where $\mathbf{1}$ is the identity matrix of order 2. Consequently, we obtain the system

$$G(\eta)\cdot\begin{vmatrix}E_{a+}(\eta)\\E_{b+}(\eta)\end{vmatrix}=\begin{vmatrix}X_-(\eta)\\Y_-(\eta)\end{vmatrix} \quad (4.3.1.46)$$

which is equivalent to the system (4.3.1.18)

$$G_i(\eta)\cdot\begin{vmatrix}X_-(\eta)\\Y_-(\eta)\end{vmatrix}=\begin{vmatrix}E_{a+}(\eta)\\E_{b+}(\eta)\end{vmatrix} \quad (4.3.1.47)$$

where $G_i(\eta)=G^{-1}(\eta)$, see (4.3.1.45).

4.3.2. *Exact solutions of the diffraction by an impedance half-plane*

4.3.2.1. *Exact solution in the presence of the isotropic impedances Z_a and Z_b*

The diffraction at skew incidence ($\alpha_o \neq 0$) by a half-plane having an isotropic surface impedance was solved by Bucci and Franceschetti (1976) using the SM method, as well as by Senior (1978) and Lüneburg and Serbest (2000), using the WH technique. In particular, the WH solution described by Lüneburg and Serbest (2000) is very interesting. In fact, introduced in this work is the important polynomial matrix (4.3.2.1) that reduces the order of the WH system from 4 to 2

$$t(\eta)=\begin{vmatrix}\eta & -\alpha_o\\ \alpha_o & \eta\end{vmatrix} \quad (4.3.2.1)$$

To solve this problem, in this book, we consider the factorization of the matrix (4.3.1.19)

$$\mathbf{G}_\mathrm{i}(\eta)=\begin{vmatrix}\frac{1}{2}\{\mathbf{1}+\mathbf{Z}_\mathrm{a}[\mathbf{Z}_\mathrm{c}(\eta)]^{-1}\} & \frac{1}{2}\{\mathbf{Z}_\mathrm{a}+\mathbf{Z}_\mathrm{c}(\eta)\}\\ -\frac{1}{2}\{\mathbf{1}+\mathbf{Z}_\mathrm{b}[\mathbf{Z}_\mathrm{c}(\eta)]^{-1}\} & \frac{1}{2}\{\mathbf{Z}_\mathrm{b}+\mathbf{Z}_\mathrm{c}(\eta)\}\end{vmatrix} \quad (4.3.2.2)$$

when the two surface impedances are scalars, i.e. $\mathbf{Z}_\mathrm{a}=Z_a\mathbf{1}$ and $\mathbf{Z}_b=Z_b\mathbf{1}$.

The key point to obtain this factorization is that the order of matrices is reduced by noting that the polynomial matrix $t(\eta)=\begin{vmatrix}\eta & -\alpha_o\\ \alpha_o & \eta\end{vmatrix}$ makes $\mathrm{Z_c}$ and $\mathrm{Y_c}$ diagonal.

In fact

$$\mathbf{Z}_c = t^{-1}(\eta)\cdot\begin{vmatrix}\frac{\xi}{kY_o} & 0\\ 0 & \frac{kZ_o}{\xi}\end{vmatrix}\cdot t(\eta), \qquad \mathbf{Y}_c = t^{-1}(\eta)\cdot\begin{vmatrix}\frac{kY_o}{\xi} & 0\\ 0 & \frac{\xi}{kZ_o}\end{vmatrix}\cdot t(\eta) \tag{4.3.2.3}$$

or

$$\begin{vmatrix}k^2-\eta^2 & \eta\,\alpha_o\\ \eta\,\alpha_o & \tau_o^2\end{vmatrix} = t^{-1}(\eta)\cdot d(\eta)\cdot t(\eta) \tag{4.3.2.4}$$

where:

$$d(\eta) = \begin{vmatrix}\xi^2 & 0\\ 0 & k^2\end{vmatrix} \tag{4.3.2.5}$$

Putting $T(\eta) = \begin{vmatrix}t(\eta) & 0\\ 0 & t(\eta)\end{vmatrix}$, it follows that

$$T(\eta)\cdot G_i(\eta)\cdot T^{-1}(\eta) = \begin{vmatrix}x & 0 & x & 0\\ 0 & x & 0 & x\\ x & 0 & x & 0\\ 0 & x & 0 & x\end{vmatrix} \tag{4.3.2.6}$$

where x is a non-null element. By introducing the matrix P defined by:

$$P = \begin{vmatrix}1 & 0 & 0 & 0\\ 0 & 0 & 1 & 0\\ 0 & 1 & 0 & 0\\ 0 & 0 & 0 & 1\end{vmatrix} \tag{4.3.2.7}$$

we can reorder to have the following quasi-diagonal matrix:

$$G_t(\eta) = P\cdot T(\eta)\cdot G_i(\eta)\cdot T^{-1}(\eta)\cdot P = \begin{vmatrix}G_{t1}(\eta) & 0\\ 0 & G_{t2}(\eta)\end{vmatrix} \tag{4.3.2.8}$$

where the matrices of order 2 $G_{t1}(\eta)$ and $G_{t2}(\eta)$ are defined by:

$$G_{t1}(\eta)=\frac{1}{2}\begin{vmatrix} (1+\frac{Z_a\omega\varepsilon}{\xi}) & \dfrac{Z_a+\frac{\xi}{\omega\varepsilon}}{Z_o} \\ -(1+\frac{Z_b\omega\varepsilon}{\xi}) & \dfrac{Z_b+\frac{\xi}{\omega\varepsilon}}{Z_o} \end{vmatrix},\quad G_{t2}(\eta)=\frac{1}{2}\begin{vmatrix} (1+\frac{Z_a\omega\varepsilon\xi}{k^2}) & \dfrac{Z_a+\frac{k^2}{\omega\varepsilon\xi}}{Z_o} \\ -(1+\frac{Z_b\omega\varepsilon\xi}{k^2}) & \dfrac{Z_b+\frac{k^2}{\omega\varepsilon\xi}}{Z_o} \end{vmatrix} \qquad (4.3.2.9)$$

In the following, we will apply the property:

$$P^2=1\rightarrow P^{-1}=P \qquad (4.3.2.10)$$

where 1 is the unit matrix of order 4. We rewrite equation (4.3.2.8) in the form

$$G_i(\eta)=T^{-1}(\eta)P\,G_t(\eta)P\,T(\eta) \qquad (4.3.2.11)$$

This equation reduces the factorization of the matrix of order 4 $G_i(\eta)$ to the factorization of the matrices of order 2 $G_{t1}(\eta)=G_{t1+}(\eta)G_{t1-}(\eta)$ and $G_{t2}(\eta)=G_{t2+}(\eta)G_{t2-}(\eta)$.

We can reduce the matrices $G_{t1}(\eta)$ and $G_{t2}(\eta)$ to matrices of Daniele–Khrapkov form (1.8.1) using the following equations obtained by algebraic manipulations:

$$G_{t1}(\eta)=\frac{n_{ha}+n_{hb}+2\xi}{4}\begin{vmatrix} 1 & 1 \\ -1 & 1 \end{vmatrix}\begin{vmatrix} \frac{1}{\xi_+} & 0 \\ 0 & \frac{1}{\sqrt{\tau_o}} \end{vmatrix}\begin{vmatrix} 1 & \frac{\xi_+}{\sqrt{\tau_o}}f_h \\ \frac{\sqrt{\tau_o}}{\xi_+}f_h & 1 \end{vmatrix}\begin{vmatrix} \frac{1}{\xi_-} & 0 \\ 0 & \frac{1}{\sqrt{\tau_o}} \end{vmatrix}\begin{vmatrix} 1 & 0 \\ 0 & \frac{\tau_o}{k} \end{vmatrix} \qquad (4.3.2.12)$$

$$G_{t2}(\eta)=\frac{n_{ea}+n_{eb}+2\xi}{4}\begin{vmatrix} \frac{1}{n_{ea}} & \frac{1}{n_{ea}} \\ -\frac{1}{n_{eb}} & \frac{1}{n_{eb}} \end{vmatrix}\begin{vmatrix} \frac{1}{\sqrt{\tau_o}} & 0 \\ 0 & \frac{1}{\xi_+} \end{vmatrix}\begin{vmatrix} 1 & \frac{\sqrt{\tau_o}}{\xi_+}f_e \\ \frac{\xi_+}{\sqrt{\tau_o}}f_e & 1 \end{vmatrix}\begin{vmatrix} \frac{1}{\sqrt{\tau_o}} & 0 \\ 0 & \frac{1}{\xi_-} \end{vmatrix}\begin{vmatrix} 1 & 0 \\ 0 & \frac{k}{\tau_o} \end{vmatrix} \qquad (4.3.2.13)$$

where we have introduced the following constants and functions:

$$n_{ea} = \tau_o \sin\theta_{ea} = \frac{\omega\mu}{Z_a}, \qquad n_{eb} = \frac{\omega\mu}{Z_b} = \tau_o \sin\theta_{eb} \tag{4.3.2.14}$$

$$n_{ha} = \tau_o \sin\theta_{ha} = \omega\varepsilon Z_a, \qquad n_{hb} = \tau_o \sin\theta_{hb} = \omega\varepsilon Z_b \tag{4.3.2.15}$$

$$\xi_- = -\sqrt{\frac{\tau_o + \eta}{2}}, \qquad \xi_+ = -\sqrt{2(\tau_o - \eta)} \tag{4.3.2.16}$$

$$f_h = \frac{n_{ha} - n_{hb}}{n_{ha} + n_{hb} + 2\xi}, \qquad f_e = \frac{n_{ea} - n_{eb}}{n_{ea} + n_{eb} + 2\xi} \tag{4.3.2.17}$$

From (4.3.2.12) and (4.3.2.13), the factorization of $G_{t1}(\eta) = G_{t1+}(\eta)G_{t1-}(\eta)$ and $G_{t2}(\eta) = G_{t2+}(\eta)G_{t2-}(\eta)$ reduces to the factorization of scalars and of the central matrices

$$\begin{vmatrix} 1 & \frac{\xi_+}{\sqrt{\tau_o}} f_h \\ \frac{\sqrt{\tau_o}}{\xi_+} f_h & 1 \end{vmatrix}, \qquad \begin{vmatrix} 1 & \frac{\sqrt{\tau_o}}{\xi_+} f_e \\ \frac{\xi_+}{\sqrt{\tau_o}} f_e & 1 \end{vmatrix} \tag{4.3.2.18}$$

These matrices are of Daniele–Khrapkov (1.8.1) type because the ratio of the off-diagonal terms gives the rational function

$$\frac{\xi_+^2}{\tau_o} = \frac{2(\tau_o - \eta)}{\tau_o} \tag{4.3.2.19}$$

Taking into account (4.3.2.19) and (1.8.3) and (1.8.4) yield the factorized matrices that are regular, i.e. bounded as $\eta \to \infty$.

The factorization of the kernel G of the WH equation (4.3.1.18) written in the form

$$G(\eta) \cdot \begin{vmatrix} E_{a+}(\eta) \\ E_{b+}(\eta) \end{vmatrix} = \begin{vmatrix} X_-(\eta) \\ Y_-(\eta) \end{vmatrix} \tag{4.3.2.20}$$

is expressed by

$$G_-(\eta) = T^{-1}(\eta) \cdot P \cdot \begin{vmatrix} G_{t1-}^{-1}(\eta) & 0 \\ 0 & G_{t2-}^{-1}(\eta) \end{vmatrix} \tag{4.3.2.21}$$

$$G_+(\eta) = \begin{vmatrix} G_{t1+}^{-1}(\eta) & 0 \\ 0 & G_{t2+}^{-1}(\eta) \end{vmatrix} \cdot P \cdot T\ (\eta) \tag{4.3.2.22}$$

This factorization is weak (Daniele and Zich 2014, section 2.7) for the presence of offending poles that arise from $T^{-1}(\eta)$. These poles are:

$$\eta = \pm \eta_f \tag{4.3.2.23}$$

where $\eta_f = j\alpha_o$ ($\alpha_o = k \sin\beta$).

To overcome this problem, by assuming the presence of a non-standard pole η_o of the plus unknown

$$F_+(\eta) = \begin{vmatrix} E_{a+}(\eta) \\ E_{b+}(\eta) \end{vmatrix} \tag{4.3.2.24}$$

we use (1.5.6):

$$G_+(\eta)F_+(\eta) - G_+(\eta_o)F_+^{n.s.}(\eta) = [G_-(\eta)]^{-1} F_-(\eta) - G_+(\eta_o)F_+^{n.s.}(\eta) = w(\eta) = |w_1 \quad w_2 \quad w_3 \quad w_4|^t \tag{4.3.2.25}$$

where by definitions (4.3.1.10) and (4.3.1.11)

$$F_-(\eta) = \begin{vmatrix} X_-(\eta) \\ Y_-(\eta) \end{vmatrix}, \quad F_+^{n.s}(\eta) = \begin{vmatrix} E_{a+}^{n.s.}(\eta) \\ E_{b+}^{n.s.}(\eta) \end{vmatrix} = \begin{vmatrix} E_{a+}^{GO.}(\eta) \\ E_{b+}^{GO}(\eta) \end{vmatrix} \tag{4.3.2.26}$$

with

$$E_{a+}^{GO.}(\eta) = \begin{vmatrix} -V_{\rho+}^{i}(\eta,0) \\ V_{z+}^{i}(\eta,0) \end{vmatrix} + Z_a \begin{vmatrix} I_{z+}^{i}(\eta,0) \\ I_{\rho+}^{i}(\eta,0) \end{vmatrix}, \quad E_{b+}^{GO.}(\eta) = \begin{vmatrix} -V_{\rho+}^{i}(\eta,0) \\ V_{z+}^{i}(\eta,0) \end{vmatrix} - Z_b \begin{vmatrix} I_{z+}^{i}(\eta,0) \\ I_{\rho+}^{i}(\eta,0) \end{vmatrix} \tag{4.3.2.27}$$

Now note that considering the behavior of $G_+(\eta)$ as $\eta \to \infty$, the Liouville theorem imposes that the entire function $w(\eta) = w$ is a non-vanishing unknown constant. We evaluate $w(\eta) = w$ using the following considerations. First, we can express the plus and minus WH unknowns by equations (4.3.2.25) that can be rewritten as

$$\begin{vmatrix} E_{a+}(\eta) \\ E_{b+}(\eta) \end{vmatrix} = \frac{T_a(\eta)}{\eta^2+\alpha_o^2} \cdot P \cdot G_{t+}(\eta)\left[w(\eta) + G_+(\eta_o) F_+^{n.s.}(\eta) \right] \tag{4.3.2.28}$$

$$\begin{vmatrix} X_-(\eta) \\ Y_-(\eta) \end{vmatrix} = \frac{T_a(\eta)}{\eta^2+\alpha_o^2} \cdot P \cdot G_{t-}^{-1}(\eta)\left[w(\eta) + G_+(\eta_o) F_+^{n.s.}(\eta) \right] \tag{4.3.2.29}$$

where:

$$T_a(\eta) = \begin{vmatrix} t_o(\eta) & 0 \\ 0 & t_o(\eta) \end{vmatrix}, \qquad t_o(\eta) = \begin{vmatrix} \eta & \alpha_o \\ -\alpha_o & \eta \end{vmatrix}. \tag{4.3.2.30}$$

with $t_o(\eta)$ defined as the adjoint of (4.3.2.1).

Expression (4.3.2.28) apparently contains the pole $\eta_f = j\alpha_o$ ($\alpha_o = k\sin\beta$) at the second member. Similarly, (4.3.2.29) apparently contains the offending pole $-\eta_f = -j\alpha_o$ at the second member. To avoid these offending behaviors, the vectors $P \cdot G_{t+}(\eta_f)\left[w + G_+(\eta_o) F_+^{n.s.}(\eta_f) \right]$ and $P \cdot G_{t-}^{-1}(-\eta_f)\left[w + G_+(\eta_o) F_+^{n.s}(-\eta_f) \right]$ must be respectively in the null spaces of $T_a(\eta_f)$ and $T_a(-\eta_f)$.

The null space of $T_a(\eta_f)$ is given by the two vectors $u_1 = |0 \;\; 0 \;\; j \;\; 1|^t$, $u_2 = |j \;\; 1 \;\; 0 \;\; 0|^t$ and the one of $T_a(-\eta_f)$ is given by the two vectors $u_3 = |0 \;\; 0 \;\; -j \;\; 1|^t$, $u_4 = |-j \;\; 1 \;\; 0 \;\; 0|^t$.

The definition of null space impose the eight scalar equations

$$P \cdot G_{t+}(\eta_f)\left[w + G_+(\eta_o) F_+^{n.s.}(\eta_f) \right] = C_1 u_1 + C_2 u_2 \tag{4.3.2.31}$$

$$P \cdot G_{t-}^{-1}(-\eta_f)\left[w + G_+(\eta_o) F_+^{n.s.}(-\eta_f) \right] = C_3 u_3 + C_4 u_4 \tag{4.3.2.32}$$

in the eight scalar unknowns $w_1, w_2, w_3, w_4, C_1, C_2, C_3, C_4$.

The solution of (4.3.2.31) and (4.3.2.32) provides the knowledge of w and consequently using (4.3.2.28) and (4.3.2.29) we obtain the WH spectra $\begin{vmatrix} E_{a+}(\eta) \\ E_{b+}(\eta) \end{vmatrix}$ and $F_-(\eta) = \begin{vmatrix} X_-(\eta) \\ Y_-(\eta) \end{vmatrix}$.

4.3.2.2. *Exact solution in the presence of symmetric anisotropic impedances $Z_a = Z_b$*

In order to show the complexity of obtaining exact factorization even when they are possible, in this section, we consider the explicit factorization of the WH matrix for a symmetric anisotropic half-plane $\mathbf{Z_a} = \mathbf{Z}_b$. The WH solution for the diffraction by an impedance problem with diagonal $\mathbf{Z_a} = \mathbf{Z}_b$ was obtained by Hurd and Lüneburg (1985). The solution presented here also holds in the more general case of arbitrary $\mathbf{Z_a} = \mathbf{Z}_b$. Considering (4.3.1.22), the factorization of the relevant matrix of order 4 is reduced to the factorization of the matrix $\mathbf{Z_a} + \mathbf{Z}_c(\eta)$ of order 2, which can be accomplished by noting that

$$1 + \frac{1}{\xi}(\mathbf{Z_a})^{-1}(\xi\,\mathbf{Z}_c(\eta)) \tag{4.3.2.33}$$

commutes with the polynomial matrix $P_2(\eta) = (\mathbf{Z_a})^{-1}(\xi\,\mathbf{Z}_c(\boldsymbol{\eta}))$. Note that (4.3.2.33) differs from $\mathbf{Z_a} + \mathbf{Z}_c(\eta)$ of a constant matrix; therefore, they have the same factorizations.

The logarithmic decomposition yields a non-standard factorization

$$W = 1 + \frac{1}{\xi}(\mathbf{Z_a})^{-1}(\xi\,\mathbf{Z}_c(\eta)) = \tilde{W}_+(\eta)\tilde{W}_-(\eta) \tag{4.3.2.34}$$

where $\tilde{W}_+(\eta)$ and $\tilde{W}_-(\eta)$ show unbounded behavior at infinity. To eliminate this behavior at infinity we use the general procedure described in Daniele and Zich (2014), which consists of considering the properties of the polynomial matrix $P_2(\eta)$. Since the polynomial matrix $P_2(\eta)$ is of order 2, it is relatively easy to eliminate the offending exponential behavior of the factorized matrices (Daniele 1984b). Considering the general procedure described in section 4.9.2 of Daniele and Zich (2014), we introduce the rational matrix $R(\eta) = 1 + xP_2(\eta)$ and its factorization. To eliminate the offending exponential behaviors of these factorized matrices, we need to find x such that it satisfies the nonlinear equation:

$$f(x)=\int_{-\infty}^{\infty}\left(\frac{\log[1+x\lambda_1]}{\lambda_1-\lambda_2}+\frac{\log[1+x\lambda_2]}{\lambda_2-\lambda_1}\right)d\eta=n \tag{4.3.2.35}$$

where $\lambda_1=\lambda_1(\eta)$ and $\lambda_2=\lambda_2(\eta)$ are the two eigenvalues of $P_2(\eta)$ and n is a known quantity defined by the integral

$$n=\int_{-\infty}^{\infty}\left(\frac{\log[1+\frac{1}{\xi(\eta)}\lambda_1]}{\lambda_1-\lambda_2}+\frac{\log[1+\frac{1}{\xi(\eta)}\lambda_2]}{\lambda_2-\lambda_1}\right)d\eta\,.$$

From (4.3.2.35), we have

$$\frac{df(x)}{dx}=\int_{-\infty}^{\infty}\left(\frac{1}{1+x(\lambda_1+\lambda_2)+x^2\lambda_2\lambda_1}\right)d\eta \tag{4.3.2.36}$$

The integrand of (4.3.2.36) is rational in η. By applying the residue theorem, it yields

$$\frac{df(x)}{dx}=2\pi j\frac{k^2}{\sqrt{x(ax^3+bx^2+cx+d)}} \tag{4.3.2.37}$$

where:

$$a=4\det^2[z_a]\tau_o^2$$

$$b=4\det[z_a][k^2z_{11}+(z_{11}+z_{22})\tau_o^2]$$

$$c=4k^2(z_{11}^2+\det[z_a])+(z_{12}+z_{21})^2\alpha_o^2+4z_{11}z_{22}\tau_o^2$$

$$d=4k^2z_{11} \tag{4.3.2.38}$$

and z_{ij} are the normalized values of the impedance $\mathbf{Z_a}=\mathbf{Z}_b$. Taking into account that $f(0)=0$, integrating (4.3.2.36), we obtain $f(x)$ expressed in terms of elliptical integrals:

$$f(x)=\frac{1}{\sqrt{x_2x_1(x_1-x_3)}}F(\arcsin[\sqrt{\frac{x(x_3-x_1)}{(x-x_1)x_3}},\frac{x_1(x_1-x_2)}{(x_1-x_3)x_2})=n \tag{4.3.2.39}$$

where $F(\phi,m)$ is the elliptical integral of the first kind and the zeros of $ax^3+bx^2+cx+d=a(x-x_1)(x-x_2)(x-x_3)$ are introduced.

By inverting this equation, we explicitly express the solution x of nonlinear equations $f(x)=n$ in terms of the Jacobian elliptical function $sn(u,m)$:

$$x=\frac{x_1x_3sn[\frac{1}{2}(n x_1^2\sqrt{x_2}-n x_1 x_3\sqrt{x_2},\frac{x_3(x_1-x_2)}{(x_1-x_3)x_2}]^2}{x_1-x_3+x_3sn[\frac{1}{2}(n x_1^2\sqrt{x_2}-n x_1 x_3\sqrt{x_2},\frac{x_3(x_1-x_2)}{(x_1-x_3)x_2}]^2} \qquad (4.3.2.40)$$

Once x is obtained, in order to complete the factorization process, we also need the standard factorization of the rational matrix $R^{-1}(\eta)=(1+xP_2(\eta))^{-1}=[R^{-1}]_+[R^{-1}]_-$. It can be obtained by the general method of factorization of rational matrices described in section 4.4.2 of Daniele and Zich (2014).

The standard factorization of $W=1+\frac{1}{\xi}(\mathbf{Z_a})^{-1}(\xi\,\mathbf{Z}_c(\boldsymbol{\eta}))=\tilde{W}_+\tilde{W}_-=W_+W_-$ follows from

$$1+\frac{1}{\xi}(\mathbf{Z_a})^{-1}(\xi\,\mathbf{Z}_c(\eta))=\tilde{W}_+\tilde{R}_+R^{-1}\tilde{R}_-\tilde{W}_-=\tilde{W}_+\tilde{R}_+[R^{-1}]_+[R^{-1}]_-\tilde{R}_-\tilde{W}_- \qquad (4.3.2.41)$$

It yields the regular factorized matrices of W:

$$W_+=\tilde{W}_+\tilde{R}_+[R^{-1}]_+, \qquad W_-=[R^{-1}]_-\tilde{R}_-\tilde{W}_- \qquad (4.3.2.42)$$

Interestingly, for this problem, the SM method yields a difference equation of order 2 (Senior and Legault 1998). In general, we state that the WH factorization of matrices of order 2 yields to difference equations of order 2 in the w plane in terms of axial spectra and rotating waves. It is possible to demonstrate this property with the help of the theory reported in Chapter 3.

4.3.2.3. *Exact solution in the presence of opposite diagonal impedances $Z_b=-Z_a$ at skew incidence*

In this case, the matrix kernel (4.3.1.22) reduces to:

$$m = \begin{vmatrix} \mathbf{Z}_a & \xi_+ \mathbf{Z}_c(\eta) \\ \frac{1}{\xi_+} \mathbf{Z}_c(\eta) & \mathbf{Z}_a \end{vmatrix} = \begin{vmatrix} \mathbf{Z}_a & \frac{1}{\xi_-} \mathbf{P}(\eta) \\ \frac{1}{\xi_+^2 \xi_-} \mathbf{P}(\eta) & \mathbf{Z}_a \end{vmatrix} = M_+ M_- \qquad (4.3.2.43)$$

where the polynomial matrix $\mathbf{P}(\eta) = \xi \mathbf{Z}_c(\eta)$ is of order 2. The weak factorization of this matrix is trivial since it is already available inside its definition, i.e. it is a weak minus matrix where the offending pole results from $\frac{1}{\xi_+^2} = \frac{1}{\tau_o - \eta}$. Consequently, the weak factorization is $m = m_{w+} m_{w-}$ with $m_{w+} = 1$ and $m_{w-} = m$. The procedure for making any weak factorization regular is described in section 2.7 of Daniele and Zich (2014).

4.3.2.4. *Exact solution in the presence of non-invertible Z_a ($det[Z_a] = 0$) and Z_b ($det[Z_b] = 0$) at skew incidence*

In general, if a matrix $\mathbf{Z}$ of order 2 has a null determinant $\det[\mathbf{Z}] = 0$, we can define three parameters Z_f, δ and p that allow the following representation:

$$Z = \begin{vmatrix} Z_{11} & Z_{12} \\ Z_{21} & Z_{22} \end{vmatrix} = Z_f \begin{vmatrix} Cos^2\delta & p\, Cos\delta\, Sin\delta \\ \frac{1}{p} Cos\delta\, Sin\delta & Sin^2\delta \end{vmatrix} \qquad (4.3.2.44)$$

where:

$$Z_f = Z_{11} + Z_{22}, \qquad \cot^2\delta = \frac{Z_{11}}{Z_{22}}, \qquad p = \sqrt{\frac{Z_{12}}{Z_{21}}} \qquad (4.3.2.45)$$

For example, in the case of reciprocity, we have $Z_{21} = Z_{12}$, which yields $p = 1$. The angle δ locates the principal anisotropy axes with respect to the Cartesian axes, where $Z = \begin{vmatrix} Z_{11} & Z_{12} \\ Z_{21} & Z_{22} \end{vmatrix}$ is defined. These axes are orthogonal if $p = 1$.

Since $\det[\mathbf{Z}]$ is null, one of the eigenvalues of $\mathbf{Z}$ must be zero. It means that a vanishing scalar surface impedance is exhibited in one of the principal anisotropy axes. Let us introduce the matrices

$$c = \begin{vmatrix} Cos^2\delta & p\,Cos\delta\,Sin\delta \\ \frac{1}{p}Cos\delta\,Sin\delta & Sin^2\delta \end{vmatrix}, \qquad c_a = \begin{vmatrix} Sin^2\delta & -p\,Cos\delta\,Sin\delta \\ -\frac{1}{p}Cos\delta\,Sin\delta & Cos^2\delta \end{vmatrix},$$

$$\sigma = \begin{vmatrix} 0 & p\,j \\ -\frac{1}{p}j & 0 \end{vmatrix} \tag{4.3.2.46}$$

The matrices c and c_a are complementary non-invertible matrices of order 2 because

$$c \cdot c_a = 0, \qquad c_a \cdot c = 0, \qquad c + c_a = \begin{vmatrix} 1 & 0 \\ 0 & 1 \end{vmatrix}, \qquad c_a^2 = c_a, \quad c^2 = c \tag{4.3.2.47}$$

In the following, we will apply the useful properties

$$\sigma\, c_a = c\sigma, \qquad \sigma\, c = c_a\,\sigma \tag{4.3.2.48}$$

where the matrix σ is invertible and satisfies the relationship: $\sigma^2 = \begin{vmatrix} 1 & 0 \\ 0 & 1 \end{vmatrix}$.

The diffraction by an anisotropic impedance face at skew incidence having a PEC face "a" and a non-invertible impedance at the face "b" with $p_b = 1$ was solved in Manara *et al.* (2004) using the SM method. In this section, we present the WH solution under less restrictive conditions:

$$Z_a = Z_{fa}c, \qquad Z_b = Z_{fb}c \tag{4.3.2.49}$$

where Z_{fa} and Z_{fb} are non-null scalar and we have arbitrary p and δ.

Let us introduce the matrices of order 4:

$$\mathbf{C} = \begin{vmatrix} c & 0 \\ 0 & c \end{vmatrix}, \ \mathbf{C}_a = \begin{vmatrix} c_a & 0 \\ 0 & c_a \end{vmatrix} \tag{4.3.2.50}$$

The kernel matrix (4.3.1.22)

$$M = \begin{vmatrix} \{Z_a - Z_b\} & \xi_+ (Z_a + Z_b + 2Z_c(\eta)) \\ \frac{1}{\xi_+}(Z_a + Z_b + 2Z_c(\eta)) & \{Z_a - Z_b\} \end{vmatrix} \tag{4.3.2.51}$$

of the WH equation

$$M\, x_- = x_+ \tag{4.3.2.52}$$

can be rewritten as

$$M = C \cdot M \cdot C + C \cdot M \cdot C_a + C_a \cdot M \tag{4.3.2.53}$$

where the elements of the matrix M of order 4 are written in terms of the sub-matrix of order 2

$$c \cdot Z_c(\eta) \cdot c\,, \qquad c \cdot Z_c(\eta) \cdot c_a\,, \qquad c_a \cdot Z_c(\eta) \cdot c\,, \quad c_a \cdot Z_c(\eta) \cdot c_a \tag{4.3.2.54}$$

defined by (4.3.2.50), (4.3.2.51) and (4.3.2.53). Taking into account (4.3.1.14), the direct calculation yields:

$$c \cdot Z_c(\eta) \cdot c = f_1(\eta)\ c\,, \qquad c \cdot Z_c(\eta) \cdot c_a = f_2(\eta)\ \sigma \cdot c_a\,, \tag{4.3.2.55}$$

$$c_a \cdot Z_c(\eta) \cdot c = -f_4(\eta)\ \sigma \cdot c\,, \qquad c_a \cdot Z_c(\eta) \cdot c_a = f_3(\eta)\ c_a \tag{4.3.2.56}$$

where the functions $f_i(\eta)$ ($i = 1,2,3,4$) are scalars. Here, we show their expressions in the case of p=1:

$$\begin{cases} f_1(\eta) = \dfrac{(k^2 - \eta^2)\cos^2\delta + \tau_o^2 \sin^2\delta + \alpha_o \eta \sin(2\delta)}{\varepsilon\omega\xi} \\ f_2(\eta) = f_4(\eta) = -j\dfrac{(-k^2 + \eta^2 + \tau_o^2)\sin(2\delta) + 2\alpha_o \eta \cos(2\delta)}{2\varepsilon\omega\xi} \\ f_3(\eta) = \dfrac{(k^2 - \eta^2)\sin^2\delta + \tau_o^2 \cos^2\delta - \alpha_o \eta \sin(2\delta)}{\varepsilon\omega\xi} \end{cases} \tag{4.3.2.57}$$

We define the solution in the form

$$x_- = C\ x_- + C_a\, x_- \tag{4.3.2.58}$$

In order to obtain an equation in Cx_-, let us multiply (4.3.2.52) by C. Taking into account (4.3.2.55), (4.3.2.56) and (4.3.2.58), it yields:

$$\begin{vmatrix} \{Z_{fa} - Z_{fb}\} & \xi_+[Z_{fa} + Z_{fb} + 2f_1(\eta)] \\ \frac{1}{\xi_+}[Z_{fa} + Z_{fb} + 2f_1(\eta)] & \{Z_{fa} - Z_{fb}\} \end{vmatrix} Cx_- + \begin{vmatrix} 0 & \xi_+ 2f_2(\eta) \\ \frac{1}{\xi_+} 2f_2(\eta) & 0 \end{vmatrix} \sigma C_a x_- = Cx_+$$

(4.3.2.59)

To obtain a link between Cx_- and $\sigma C_a x_-$, let us multiply (4.3.2.52) by C_a. We then obtain:

$$C_a M = 2\begin{vmatrix} 0 & \xi_+ c_a \mathbf{Z}_c(\eta) \\ \frac{1}{\xi_+} c_a \mathbf{Z}_c(\eta) & 0 \end{vmatrix} = 2\begin{vmatrix} 0 & \frac{\xi_+}{\xi} \\ \frac{1}{\xi_+ \xi} & 0 \end{vmatrix} \begin{vmatrix} c_a P(\eta) & 0 \\ 0 & c_a P(\eta) \end{vmatrix} \qquad (4.3.2.60)$$

with the polynomial matrix $\mathrm{P}(\eta) = \xi\, \mathrm{Z}_c(\eta)$. Since the explicit factorization of

$$\begin{vmatrix} 0 & \frac{\xi_+}{\xi} \\ \frac{1}{\xi_+ \xi} & 0 \end{vmatrix} \qquad (4.3.2.61)$$

can be accomplished in closed form, from the above equation, we obtain:

$$C_a P(\eta) x_- = n(\eta) \qquad (4.3.2.62)$$

where $n(\eta)$ is known (except for a finite number of constants derived from the presence of unbounded minus functions). Substituting (4.3.2.58), it yields

$$C_a P C_a x_- + C_a P C\, x_- = n(\eta) \qquad (4.3.2.63)$$

and taking into account (4.3.2.55) and (4.3.2.56), we obtain the required relation between Cx_- and $C_a x_-$:

$$\xi f_3(\eta) C_a x_- - \xi f_4(\eta) \sigma C\, x_- = n(\eta) \qquad (4.3.2.64)$$

From the above equation, we obtain

$$C_a x_- = \frac{f_4(\eta)}{f_3(\eta)} \sigma C\, x_- + \frac{n(\eta)}{\xi f_3(\eta)} \tag{4.3.2.65}$$

or

$$\sigma C_a x_- = \frac{f_4(\eta)}{f_3(\eta)} C\, x_- + \frac{\sigma n(\eta)}{\xi f_3(\eta)} \tag{4.3.2.66}$$

Substituting into (4.3.2.59), we need to factorize the matrix

$$M_1 = \begin{vmatrix} \{Z_{fa} - Z_{fb}\} & \xi_+ [Z_{fa} + Z_{fb} + 2(f_1(\eta) + \frac{f_2(\eta) f_4(\eta)}{f_3(\eta)})] \\ \frac{1}{\xi_+}[Z_{fa} + Z_{fb} + 2(f_1(\eta) + \frac{f_2(\eta) f_4(\eta)}{f_3(\eta)})] & \{Z_{fa} - Z_{fb}\} \end{vmatrix} \tag{4.3.2.67}$$

which is of order 2. Since it has the Daniele–Khrapkov form, it can be explicitly factorized. The solution of the WH equation:

$$M_1 C\, x_- = C\, x_+ + N(\eta) \tag{4.3.2.68}$$

where $N(\eta)$ is related to $n(\eta)$, provides the unknown $C\, x_-$. Once $C\, x_-$ is obtained, $C_a\, x_-$ can be derived from (4.3.2.66). Finally, we complete the solutions $x_- = C\ x_- + C_a\, x_-$ and $x_+ = M\ x_-$ in terms of $n(\eta)$. To evaluate the finite number of constants present in $n(\eta)$, we need to impose the condition that $x_-(\eta)$ and $x_+(\eta)$ are Laplace transforms with non-offensive poles in their regularity half-planes (see weak factorization in Daniele and Zich (2014, section 2.7) already applied in section 4.3.2 of this book). These poles derive from the zeros of the determinant of M_1:

$$Det[M_1] = -4\,[Z_{fa} + f_1(\eta) + \frac{f_2(\eta) f_4(\eta)}{f_3(\eta)}][Z_{fb} + f_1(\eta) + \frac{f_2(\eta) f_4(\eta)}{f_3(\eta)}] \tag{4.3.2.69}$$

4.3.2.5. *Exact solution for the diffraction by a half-plane at normal incidence with diagonal impedances*

We recall that if **f** is diagonal in (4.3.1.24), the matrix $\mathbf{M}(\eta)$ can be explicitly factorized. In order to have a diagonal **f**, it is sufficient that $\mathbf{\Gamma}_i$, Y_Δ and $\mathrm{Z}_c(\eta)$ are diagonal. Considering the expressions of these matrices, we note that a diagonal

$Z_c(\eta)$ implies normal incidence ($\alpha_o = 0$). Diagonal $\boldsymbol{\Gamma}_i$ and Y_Δ implies diagonal $\mathbf{Z}_a$ and $\mathbf{Z}_b$. Thus, the diffraction by a half-plane at normal incidence with diagonal impedances has a closed-form solution.

4.3.2.6. *Exact factorization with the scalar reflection coefficient* $\boldsymbol{\Gamma} = (\boldsymbol{Z}_a + \boldsymbol{Z}_b)^{-1}(\boldsymbol{Z}_a - \boldsymbol{Z}_b)$

We recall that since in (4.3.1.24) we deal with a matrix that commutes with a polynomial matrix, we can obtain an exact factorization of matrix (4.3.1.24) using the general procedure described in Daniele and Zich (2014). In this section, we show that this method is not excessively cumbersome when the ratio of the two face impedances is a scalar or, more generally, their reflection coefficient is a scalar

$$\boldsymbol{\Gamma} = (\mathbf{Z}_a + \mathbf{Z}_b)^{-1}(\mathbf{Z}_a - \mathbf{Z}_b) = \frac{1}{a}\mathbf{1} \qquad (4.3.2.70)$$

However, again, we deal with a "not simple" problem since the solution of nonlinear algebraic equations is required. The above simplification yields the following form of **f** (4.3.1.24) that commutes with a polynomial matrix $\mathrm{P}(\eta)$:

$$\mathrm{f} = a1 + \frac{k}{\xi}\mathrm{P}(\eta) \qquad (4.3.2.71)$$

where:

$$\mathbf{P}(\eta) = \frac{Z_o \mathbf{Y}_\Delta}{k^2}\begin{vmatrix} k^2 - \eta^2 & \eta\,\alpha_o \\ \eta\,\alpha_o & \tau_o^2 \end{vmatrix} \qquad (4.3.2.72)$$

Here, we briefly summarize the factorizing procedure of this case. We begin with the factorization formula of section 1.8:

$$\begin{vmatrix} 1 & \xi_+ \mathbf{f} \\ \frac{1}{\xi_+}\mathbf{f} & 1 \end{vmatrix} = \tilde{\mathbf{g}}_+ \begin{vmatrix} \cosh\left(\frac{1}{2}\xi_+ \mathbf{t}_+\right) & \xi_+ \sinh\left(\frac{1}{2}\xi_+ \mathbf{t}_+\right) \\ \frac{1}{\xi_+}\sinh\left(\frac{1}{2}\xi_+ \mathbf{t}_+\right) & \cosh\left(\frac{1}{2}\xi_+ \mathbf{t}_+\right) \end{vmatrix} \cdot \begin{vmatrix} \cosh\left(\frac{1}{2}\xi_+ \mathbf{t}_-\right) & \xi_+ \sinh\left(\frac{1}{2}\xi_+ \mathbf{t}_-\right) \\ \frac{1}{\xi_+}\sinh\left(\frac{1}{2}\xi_+ \mathbf{t}_-\right) & \cosh\left(\frac{1}{2}\xi_+ t_-\right) \end{vmatrix} \tilde{\mathbf{g}}_- \qquad (4.3.2.73)$$

where a commutative algebra is implicit because all the matrices of order 2 commute with the polynomial matrix $\mathbf{P}(\eta)$. Henceforth, these matrices will be indicated as scalars.

In (4.3.2.73), the following decomposition and notations are used:

$$g = \sqrt{1-\mathbf{f}^2} = \tilde{g}_- \tilde{g}_+$$
$$Log[g] = \psi_o 1 + \psi_1 P(\eta) = \psi_- + \psi_+$$
$$t = \frac{1}{\xi_+}\log\frac{1+f}{1-f} = t_o 1 + t_1 P(\eta) = t_- + t_+$$

$$\psi_\pm = \{\psi_o\}_\pm 1 + \{\psi_1\}_\pm P(\eta)$$

$$t_\pm = \{t_o\}_\pm 1 + \{t_1\}_\pm P(\eta)$$

$$\tilde{g}_\pm = \exp[\psi_\pm] \tag{4.3.2.74}$$

with

$$\psi_o(\alpha) = \frac{\log[\sqrt{1-(a+\frac{k}{\xi}\lambda_2)^2}]\lambda_1 - \log[\sqrt{1-(a+\frac{k}{\xi}\lambda_1)^2}]\lambda_2}{\lambda_1-\lambda_2},$$

$$\psi_1(\alpha) = \frac{\log[\sqrt{1-(a+\frac{k}{\xi}\lambda_1)^2}] - \log[\sqrt{1-(a+\frac{k}{\xi}\lambda_2)^2}]}{\lambda_1-\lambda_2},$$

$$t_o(\alpha) = \frac{1}{\xi_+}\frac{\log[\dfrac{1+(a+\frac{k}{\xi}\lambda_2)}{1-(a+\frac{k}{\xi}\lambda_2)}]\lambda_1 - \log[\dfrac{1+(a+\frac{k}{\xi}\lambda_1)}{1+(a-\frac{k}{\xi}\lambda_1)}]\lambda_2}{\lambda_1-\lambda_2},$$

$$t_1(\alpha) = \frac{1}{\xi_+}\frac{\log[\dfrac{1+(a+\frac{k}{\xi}\lambda_1)}{1-(a+\frac{k}{\xi}\lambda_1)}] - \log[\dfrac{1+(a+\frac{k}{\xi}\lambda_2)}{1-(a+\frac{k}{\xi}\lambda_2)}]}{\lambda_1-\lambda_2} \tag{4.3.2.75}$$

where λ_1 and λ_2 are the two eigenvalues of $\mathbf{P}(\eta)$ (4.3.2.72).

The expressions of the factorized matrices are

$$M_{\pm} = e^{\psi_{o\pm}1+\psi_{1\pm}P} \begin{vmatrix} \cosh\left(\frac{1}{2}\xi_+ t_{\pm}\right) & \xi_+ \sinh\left(\frac{1}{2}\xi_+ t_{\pm}\right) \\ \frac{1}{\xi_+}\sinh\left(\frac{1}{2}\xi_+ t_{\pm}\right) & \cosh\left(\frac{1}{2}\xi_+ t_{\pm}\right) \end{vmatrix} \tag{4.3.2.76}$$

Since the two factors at the second member commute, we separately eliminate the offending behaviors on the factorized matrices $\tilde{g}_{\pm} = \exp[\psi_{\pm}]$ from that of the matrix

$$M_{1\pm} = \begin{vmatrix} \cosh\left(\frac{1}{2}\xi_+ t_{\pm}\right) & \xi_+ \sinh\left(\frac{1}{2}\xi_+ t_{\pm}\right) \\ \frac{1}{\xi_+}\sinh\left(\frac{1}{2}\xi_+ t_{\pm}\right) & \cosh\left(\frac{1}{2}\xi_+ t_{\pm}\right) \end{vmatrix} \tag{4.3.2.77}$$

Note that in the above expressions the terms that yield offending behaviors are related to the exponentials of $\psi_{1\pm}$ and $M_{1\pm}$.The elimination of the offending behavior of $e^{\psi_{o\pm}1+\psi_{1\pm}P}$ has been considered in section 4.9.2 of Daniele and Zich (2014). It remains to consider the regularization of the matrix $M_{1\pm}$.

In addition to vanishing terms, the argument of hyperbolic functions involves the terms: $\xi_+ t_{1\pm}(\eta)P(\eta) = \mp c_1\eta^{3/2} + \mp c_2\eta^{1/2} + O[\eta^{-1/2}]$ as $\eta \to \infty$.

The non-vanishing values of

$$c_1 = \frac{1}{2\pi j}\int_{-\infty}^{\infty} t_1(\eta)d\eta \text{ and } c_2 = \frac{1}{2\pi j}\int_{-\infty}^{\infty} \eta t_1(\eta)d\eta \tag{4.3.2.78}$$

produce exponential behaviors that we need to eliminate.

To this end, according to the procedure reported in Chapter 4 of Daniele and Zich, (2014) we multiply M_1 for the commuting rational matrices R_x, R_y

$$R_x = \begin{vmatrix} 1 & \xi_+ \ \xi_+ \frac{1}{x} P(\eta) \\ \frac{1}{\xi_+}\xi_+ \frac{1}{x} P(\eta) & 1 \end{vmatrix}, \quad R_y = \begin{vmatrix} 1 & \xi_+ \ \xi_+ \frac{1}{y} P(\eta) \\ \frac{1}{\xi_+}\xi_+ \frac{1}{y} P(\eta) & 1 \end{vmatrix}, \tag{4.3.2.79}$$

where the scalars x and y are arbitrary for the moment.

The factorization formulas of section (1.8) applied to R_x and R_y produce unbounded behavior defined by the coefficients c_{Rx1}, c_{Rx2} and c_{Ry1}, c_{Ry2} similar to c_1, c_2. The unbounded behavior of R_x and R_y can compensate the unbounded behavior of M_1 provided x and y are chosen to solve the following nonlinear equations:

$$c_1 + c_{Rx1} + c_{Ry1} = 0, \qquad c_2 + c_{Rx2} + c_{Ry2} = 0 \tag{4.3.2.80}$$

or

$$E(x) + E(y) = -c_1, \qquad F(x) + F(y) = -c_2 \tag{4.3.2.81}$$

where:

$$c_{Rx1} = \frac{1}{2\pi j}\int_{-\infty}^{\infty} \frac{1}{\xi_+} \frac{\log[\frac{x+\xi_+\lambda_1}{x-\xi_+\lambda_1}] - \log[\frac{x+\xi_+\lambda_2}{x-\xi_+\lambda_2}]}{\lambda_1 - \lambda_2} d\eta = E(x) \tag{4.3.2.82}$$

$$c_{Rx2} = \frac{1}{2\pi j}\int \eta \frac{1}{\xi_+} \frac{\log[\frac{x+\xi_+\lambda_1}{x-\xi_+\lambda_1}] - \log[\frac{x+\xi_+\lambda_2}{x-\xi_+\lambda_2}]}{\lambda_1 - \lambda_2} d\eta = F(x) \tag{4.3.2.83}$$

$$c_{Ry1} = E(y); \qquad c_{Ry2} = F(y) \tag{4.3.2.84}$$

We note that

$$\frac{dE(x)}{dx} = -\frac{1}{2\pi j}\int_{-\infty}^{\infty} 2 \frac{(x^2 + \lambda_1\lambda_2\xi_+^2)}{(-x^2 + \lambda_1^2\xi_+^2)(-x^2 + \lambda_2^2\xi_+^2)} d\eta, \qquad E(0) = 0 \tag{4.3.2.85}$$

$$\frac{dF(x)}{dx} = -\frac{1}{2\pi j}\int_{-\infty}^{\infty} 2\eta \frac{(x^2 + \lambda_1\lambda_2\xi_+^2)}{(-x^2 + \lambda_1^2\xi_+^2)(-x^2 + \lambda_2^2\xi_+^2)} d\eta, \qquad F(0) = 0 \tag{4.3.2.86}$$

The quantities $\lambda_1 + \lambda_2$, $\lambda_1\lambda_2$ and $\lambda_1^2 + \lambda_2^2 = (\lambda_1 + \lambda_2)^2 - 2\lambda_1\lambda_2$ are polynomials of η. Consequently, the partial fraction decompositions of the rational expressions of the integrands allow us to evaluate the integrals in closed form. As with the factorization in the half-plane having equal face impedances, the successive integration with respect to x gives the expressions of $E(x)$ and $F(x)$ as functions related to elliptical integrals. Thus, we obtain x and y in terms of elliptical functions.

In particular, the "not simple" case considered in this section also comprises the one where the matrices $\mathbf{Z}_a$ and $\mathbf{Z}_b$ are proportional:

$$\mathbf{Z}_a = p\mathbf{Z}_b \tag{4.3.2.87}$$

with respect to a scalar factor p.

4.3.2.7. *Exact factorization at skew incidence when $\mathbf{Z}_a = 0$ and $\mathbf{Z}_b$ diagonal with a null element*

In Manara *et al.* (2000) and section 4.4.3 of this book, the diffraction by an impedance wedge at skew incidence with $\mathbf{Z}_a = 0$ and $\mathbf{Z}_b$ diagonal with a null element is considered.

In particular, the application of the SM method was proposed in Manara *et al.* (2000).

Being the half-plane, a particular wedge also means that the half-plane problem is amenable to the exact solution (see section 4.4.3). Let us consider the formulation (4.3.1.18) and (4.3.1.19):

$$\mathbf{G}_i(\eta)\begin{vmatrix}\mathbf{X}_-(\eta)\\ \mathbf{Y}_-(\eta)\end{vmatrix} = \begin{vmatrix}\mathbf{E}_{a+}(\eta)\\ \mathbf{E}_{b+}(\eta)\end{vmatrix}, \quad \mathbf{G}_i(\eta) = \begin{vmatrix}[\frac{1}{2}\{\mathbf{1}+\mathbf{Z}_a[\mathbf{Z}_c(\eta)]^{-1}\} & \frac{1}{2}\{\mathbf{Z}_a+\mathbf{Z}_c(\eta)\}\\ -\frac{1}{2}\{\mathbf{1}+\mathbf{Z}_b[\mathbf{Z}_c(\eta)]^{-1}\} & \frac{1}{2}\{\mathbf{Z}_b+\mathbf{Z}_c(\eta)\}\end{vmatrix} \tag{4.3.2.88}$$

Taking into account $\mathbf{Z}_a = 0$, putting $\mathbf{Z}_b = \begin{vmatrix}0 & 0\\ 0 & d\end{vmatrix}$ and introducing the matrix $\mathbf{t}$ (4.3.2.1) that makes the matrix $\mathbf{Z}_c(\eta)$ diagonal (4.3.2.3), algebraic manipulations on (4.3.1.19) yield the following matrix to be factorized:

$$\begin{vmatrix}1 & 0 & \frac{\xi}{kY_o} & 0\\ 0 & 1 & 0 & \frac{kZ_o}{\xi}\\ -1 & 0 & \frac{\xi}{kY_o} & 0\\ 0 & -\frac{kZ_o+d\xi}{kZ_o} & 0 & \frac{kZ_o+d\xi}{\xi}\end{vmatrix} \tag{4.3.2.89}$$

Using the constant matrix P defined by (4.3.2.7), the above matrix decouples into two matrices of order 2:

$$\begin{vmatrix} 1 & \frac{\xi}{kY_o} \\ -1 & \frac{\xi}{kY_o} \end{vmatrix} \text{ and } \begin{vmatrix} 1 & \frac{kZ_o}{\xi} \\ -\frac{kZ_o + d\xi}{kZ_o} & \frac{kZ_o + d\xi}{\xi} \end{vmatrix} \qquad (4.3.2.90)$$

The first matrix easily decouples into two scalar factorizations. After algebraic manipulations, the second matrix reduces to a matrix of Daniele–Khrapkov type. The exact factorization then becomes straightforward.

4.3.3. *Exact solution for the full-plane junction at skew incidence*

The problem of the two half-plane junction (Figure 4.3.3.1) has been studied extensively in the past (Sendag and Serbest 2001). In practice, to obtain the WH equations, it is sufficient to consider the angular functional equations (3.3.46) and (3.3.49), where $\gamma_a = \pi \Rightarrow m = \eta$, and to impose the boundary impedance conditions on faces a and b.

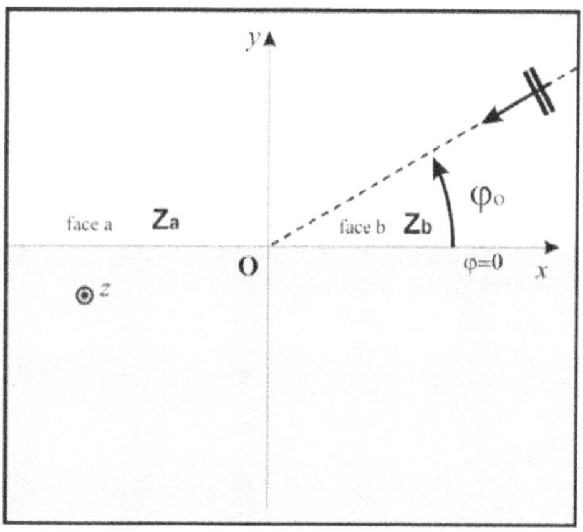

Figure 4.3.3.1. *Full-plane junction with different face impendences*

Using the same notations given in section 4.3.1, we have:

$$I_+(\eta) + I_-(\eta) = Y_c(V_+(\eta) + V_-(\eta)) \qquad (4.3.3.1)$$

$$V_+(\eta) = -Z_b I_+(\eta), \qquad V_-(\eta) = -Z_a I_-(\eta) \tag{4.3.3.2}$$

with Y_c defined in (4.3.1.14)

It yields the WH equation:

$$\left(1+Y_c Z_b\right) I_+(\eta) = -\left(1+Y_c Z_a\right) I_-(\eta) \tag{4.3.3.3}$$

If the impedances Z_b and Z_a are isotropic, the matrix

$$t(\eta) = \begin{vmatrix} \eta & -\alpha_o \\ \alpha_o & \eta \end{vmatrix} \tag{4.3.3.4}$$

makes the matrix

$$t \cdot (1+Y_c Z_a)^{-1} (1+Y_c Z_b) \cdot t^{-1} \tag{4.3.3.5}$$

diagonal and the factorization procedure becomes scalar.

At normal incidence ($\alpha_o = 0$), we also reduce to scalar factorizations the problem with diagonal Z_a and Z_b, since $(1+Y_c Z_a)^{-1}(1-Y_c Z_b)$ is diagonal.

When $\alpha_o \neq 0$, we deal with a "not simple" matrix factorization similar to section 4.3.2.6 if Z_b and Z_a are proportional. This case also includes the one where a half-plane is perfectly conducting (e.g. $Z_b = 0$) (Manara *et al.* 2000). In this case, the problem becomes simple if $\det[Z_a] = 0$, for the presence of a null eigenvalue. To show it, note that the matrix to be factorized is

$$1 + Y_c Z_a \tag{4.3.3.6}$$

For this task, we can put

$$Z_a = S \begin{vmatrix} 0 & 0 \\ 0 & d \end{vmatrix} S^{-1} \tag{4.3.3.7}$$

where d is the null eigenvalue of Z_a and S is the matrix of the eigenvectors of Z_a. Substituting (4.3.3.7) into (4.3.3.6), the factorization reduces to that of

$$1 + S^{-1} Y_C S \begin{vmatrix} 0 & 0 \\ 0 & d \end{vmatrix} \tag{4.3.3.8}$$

This matrix is triangular and can be factorized as described in section 4.3 of Daniele and Zich (2014). This simple case can also be solved using the SM method, as noted in Nepa *et al.* (2000).

4.3.4. *Exact solution of the penetrable half-plane problem (the jump)*

Using the same notation of section 4.3.1, in this case, the half-plane presents boundary conditions having the form:

$$V_{a-}(\eta) = V_{b-}(\eta) = Z_R[I_{a-}(\eta) - I_{b-}(\eta)] \tag{4.3.4.1}$$

where Z_R is an arbitrary matrix surface impedance of the penetrable half-plane (which has a continuity of the tangential electric field and a discontinuity of the tangential magnetic field).

The condition (4.3.4.1) effectively models the practical scattering problem of a penetrable half-layer with small thickness.

Taking into account that

$$V_{a+}(\eta) = V_{b+}(\eta) = V_+(\eta), \qquad I_{a+}(\eta) = I_{b+}(\eta) = I_+(\eta) \tag{4.3.4.2}$$

the WH equations of the problem can be deduced as follows:

$$V_{a-}(\eta) + V_{a+}(\eta) = Z_R[I_{a-}(\eta) - I_{b-}(\eta)] + V_+(\eta) = Z_C I_{a-}(\eta) + Z_C I_+(\eta) \tag{4.3.4.3}$$

$$V_{b-}(\eta) + V_{b+}(\eta) = Z_R[I_{a-}(\eta) - I_{b-}(\eta)] + V_+(\eta) = -Z_C I_{b-}(\eta) - Z_C I_+(\eta) \tag{4.3.4.4}$$

with Z_C defined in (4.3.1.14)

By adding and subtracting the two equations, it yields

$$2Z_R[I_{a-}(\eta) - I_{b-}(\eta)] + 2V_+(\eta) = Z_C[I_{a-}(\eta) - I_{b-}(\eta)] \tag{4.3.4.5}$$

$$0 = Z_C[I_{a-}(\eta) + I_{b-}(\eta)] + Z_C I_+(\eta) \tag{4.3.4.6}$$

or equivalently

$$2V_+(\eta) = (Z_C - 2Z_R)[I_{a-}(\eta) - I_{b-}(\eta)] \tag{4.3.4.7}$$

$$I_+(\eta) = -[I_{a-}(\eta) + I_{b-}(\eta)] \tag{4.3.4.8}$$

Equation (4.3.4.8) can be solved immediately, which yields (for plane wave sources) the incident contribution

$$I_+(\eta) = \frac{I_o^i}{\eta - \eta_o}, \qquad I_{a-}(\eta) + I_{b-}(\eta) = -\frac{I_o^i}{\eta - \eta_o} \tag{4.3.4.9}$$

In the general case, equation (4.3.4.7) involves the factorization of the kernel $Z_C - 2Z_R$. This kernel is studied in the case of the symmetric half-plane (4.3.2.2). The factorization is scalar when Z_R is a scalar. In fact, the matrix $t(\eta) = \begin{vmatrix} \eta & -\alpha_o \\ \alpha_o & \eta \end{vmatrix}$ makes the matrix $t \cdot (Z_C - 2Z_R) \cdot t^{-1}$ diagonal.

4.3.5. *Exact solution of the right-angled wedge scattering problem*

Even though it is well known that the right-angled wedge scattering problem (Figure 4.3.5.1) involves the classical WH equations, this problem is often studied using the SM method.

The WH formulation is based on (4.1.15)–(4.1.18) considered in region 1 ($\gamma_a = \pi$, $m_a = \eta$) and region 2 ($\gamma_b = \pi/2$, $m_b = \xi = \sqrt{k^2 - \eta^2}$).

By changing η with $-\eta$ in (4.1.15)–(4.1.18), we add four independent equations that introduce six new functions $V_{z+}(-\eta,0)$, $I_{z+}(-\eta,0)$, $V_{\rho+}(-\eta,0)$, $I_{\rho+}(-\eta,0)$, $I_{\rho+}(\eta,\pi)$ and $V_{\rho+}(\eta,\pi)$.

We now have eight equations that determine the 14 unknowns $V_{z+}(\eta,0)$, $I_{z+}(\eta,0)$, $V_{\rho+}(\eta,0)$, $I_{\rho+}(\eta,0)$, $I_{\rho+}(\eta,\pi)$, $V_{\rho+}(\eta,\pi)$, $I_{\rho+}(-\xi,-\frac{\pi}{2})$, $V_{\rho+}(-\xi,-\frac{\pi}{2})$, $V_{z+}(-\eta,0)$, $I_{z+}(-\eta,0)$, $V_{\rho+}(-\eta,0)$, $I_{\rho+}(-\eta,0)$, $I_{\rho+}(-\eta,\pi)$ and $V_{\rho+}(-\eta,\pi)$.

Two equations allow us to eliminate the two functions $I_{\rho+}(-\xi,-\frac{\pi}{2})$ and $V_{\rho+}(-\xi,-\frac{\pi}{2})$. Rearranging the other 12 yields a system of classical WH equations that involve the six plus functions $V_{z+}(\eta,0)$, $I_{z+}(\eta,0)$, $V_{\rho+}(\eta,0)$, $I_{\rho+}(\eta,0)$, $I_{\rho+}(\eta,\pi)$, $V_{\rho+}(\eta,\pi)$ and the six minus functions $V_{z+}(-\eta,0)$, $I_{z+}(-\eta,0)$, $V_{\rho+}(-\eta,0)$, $I_{\rho+}(-\eta,0)$, $I_{\rho+}(-\eta,\pi)$, $V_{\rho+}(-\eta,\pi)$.

Again, the kernel matrix $G(\eta)$ of this system has the form (4.1.7). Consequently, the same considerations apply as in the half-plane problem; in particular, in all the cases, we need to factorize kernels that commute with polynomial matrices. In most cases, we deal with "not simple" problems. In fact, the factorized matrices obtained with the logarithmic decomposition are almost always unbounded at infinity (Daniele and Zich 2014).

However, there are particular sets of face impedances of right-angled wedges that are solved exactly using the SM method from difference equation formulations, see section 4.3 of Senior and Volakis (1995). Hence, it is interesting to discuss these cases in the framework of WH factorization.

We show that these cases yield factorized matrices that do not contain offending singularities at infinity.

Ascertaining all the cases where an exact matrix factorization is possible is a difficult task. In this section, we limit the study to an important case that is solved using the SM method (Manara and Nepa 2000): the diffraction at skew incidence by a right-angled impedance wedge with a diagonal anisotropic impedance face, and the other perfectly electrically conducting.

The same case at normal incidence was considered in Daniele and Zich (2014). Here, we consider skew incidence with a diagonal impedance $\mathbf{Z}_a$ on face a and with an impedance $\mathbf{Z}_b = 0$ on face b (Figure 4.3.5.1)

This problem is solved using the SM method (Manara and Nepa 2000) and it is a generalization of the solution with isotropic impedance Z_a given in Vaccaro (1981), using a slight modification of the SM method.

An important observation made in Manara and Nepa (2000) is that this problem reduces to a more simple problem if we introduce a modified electromagnetic field $\breve{\mathrm{E}}, \breve{\mathrm{H}}$ related to the original one E, H by the equations:

$$\breve{E}_z = \frac{Z_o}{Z_2} E_z + Z_o H_x \tag{4.3.5.1}$$

$$\breve{H}_z = \frac{Z_1}{Z_o} H_z - \frac{1}{Z_o} E_x \tag{4.3.5.2}$$

where the diagonal anisotropic impedance Z_a is defined by the equations valid on the face a:

$$\begin{vmatrix} E_z \\ H_z \end{vmatrix} = -\begin{vmatrix} Z_2 & 0 \\ 0 & \dfrac{1}{Z_1} \end{vmatrix} \begin{vmatrix} -H_\rho \\ E_\rho \end{vmatrix} \tag{4.3.5.3}$$

that is different from the definition given in (4.3.1.9).

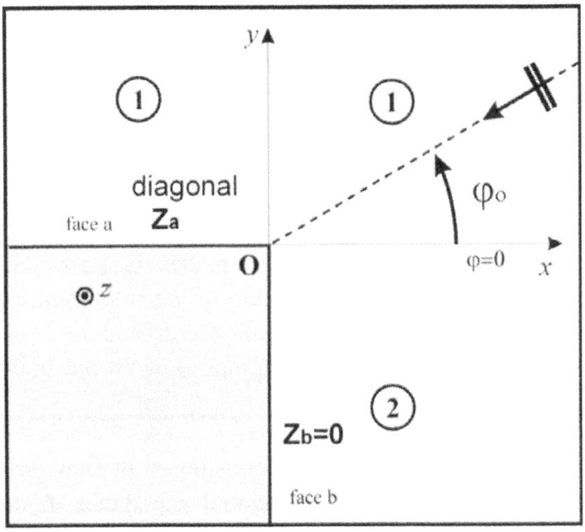

Figure 4.3.5.1. *An example of a right-angled wedge that can be solved exactly at skew incidence*

Definitions (4.3.5.1) and (4.3.5.2) satisfy the wave equations:

$$(\nabla^2 + k^2)\begin{vmatrix} \breve{E}_z \\ \breve{H}_z \end{vmatrix} = 0 \tag{4.3.5.4}$$

In addition, by defining the other field components of $\breve{\mathrm{E}}, \breve{\mathrm{H}}$ by transversal equations (see page 125 of Senior and Volakis 1995)

$$\breve{E}_\rho = \frac{1}{jk\sin^2(\beta)}\left[\cos(\beta)\frac{\partial \breve{E}_z}{\partial \rho} + Z_o\frac{1}{\rho}\frac{\partial \breve{H}_z}{\partial \varphi}\right] \tag{4.3.5.5}$$

$$\breve{E}_\varphi = \frac{1}{jk\sin^2(\beta)}\left[\cos(\beta)\frac{1}{\rho}\frac{\partial \breve{E}_z}{\partial \varphi} - Z_o\frac{\partial \breve{H}_z}{\partial \rho}\right] \tag{4.3.5.6}$$

$$Z_o\breve{H}_\rho = \frac{1}{jk\sin^2(\beta)}\left[\cos(\beta)Z_o\frac{\partial \breve{H}_z}{\partial \rho} - \frac{1}{\rho}\frac{\partial \breve{E}_z}{\partial \varphi}\right] \tag{4.3.5.7}$$

$$Z_o\breve{H}_\varphi = \frac{1}{jk\sin^2(\beta)}\left[\cos(\beta)Z_o\frac{1}{\rho}\frac{\partial \breve{H}_z}{\partial \varphi} + \frac{\partial \breve{E}_z}{\partial \rho}\right] \tag{4.3.5.8}$$

it follows that $\breve{\mathrm{E}}, \breve{\mathrm{H}}$ satisfy the Maxwell equations under the following boundary conditions on the faces of the wedge:

$$\breve{E}_z = 0\,,\ \breve{H}_z = 0 \ \text{ at } \varphi = \pi \ \text{ (face a)} \tag{4.3.5.9}$$

$$\breve{E}_z = 0\,,\ \breve{E}_\rho = 0 \ \text{ at } (\varphi = -\frac{\pi}{2}) \text{ (PEC face b)} \tag{4.3.5.10}$$

Using this formulation, we have a simplified boundary value problem (which was solved using the SM method in Manara and Nepa (2000)).

Next, we describe the solution procedure in the framework of the WH technique.

The WH formulation does not require definitions (4.3.5.1) and (4.3.5.2), but it introduces classical WH equations with kernels that can be factorized in closed form in terms of the usual logarithmic decomposition.

This factorization is simple because the factorized matrices are bounded as $\eta \to \infty$.

We can rewrite (3.3.57)–(3.3.60) in angular regions 1 and 2 by taking into account that $\gamma_a = \pi, \gamma_b = \pi/2$ that yield $m_1 = m_a = \eta,\ \ n_1 = n_a = \xi$, $m_2 = m_b = \xi,\ \ n_2 = n_b = -\eta$:

$$\xi V_{zp} - \frac{\tau_o^2}{\omega\varepsilon}I_{\rho p} - \frac{\alpha_o\eta}{\omega\varepsilon}I_{zp} = -n_1V_{za} - \frac{\tau_o^2}{\omega\varepsilon}I_{\rho a} + \frac{\alpha_o m_1}{\omega\varepsilon}I_{za} \tag{4.3.5.11}$$

$$\xi I_{zp} + \frac{\tau_o^2}{\omega\mu} V_{\rho p} + \frac{\alpha_o \eta}{\omega\mu} V_{zp} = -n_1 I_{za} + \frac{\tau_o^2}{\omega\mu} V_{\rho a} - \frac{\alpha_o m_1}{\omega\mu} V_{za} \tag{4.3.5.12}$$

$$\xi V_{zp} + \frac{\tau_o^2}{\omega\varepsilon} I_{\rho p} + \frac{\alpha_o \eta}{\omega\varepsilon} I_{zp} = -n_2 V_{zb} + \frac{\tau_o^2}{\omega\varepsilon} I_{\rho b} - \frac{\alpha_o m_2}{\omega\varepsilon} I_{zb} \tag{4.3.5.13}$$

$$\xi I_{zp} - \frac{\tau_o^2}{\omega\mu} V_{\rho p} - \frac{\alpha_o \eta}{\omega\mu} V_{zp} = -n_2 I_{zb} - \frac{\tau_o^2}{\omega\mu} V_{\rho b} + \frac{\alpha_o m_2}{\omega\mu} V_{zb} \tag{4.3.5.14}$$

where the following notations are used:

$$V_{zp} = V_{z+}(\eta, 0),\; I_{zp} = I_{z+}(\eta, 0) \tag{4.3.5.15}$$

$$V_{\rho p} = V_{\rho +}(\eta, 0),\; I_{\rho p} = I_{\rho +}(\eta, 0) \tag{4.3.5.16}$$

$$V_{za} = V_{z+}(-\eta, \pi),\; I_{za} = I_{z+}(-\eta, \pi), V_{\rho a} = V_{\rho +}(-\eta, \pi),\; I_{\rho a} = I_{\rho +}(-\eta, \pi) \tag{4.3.5.17}$$

$$V_{zb} = V_{z+}(-\xi, -\frac{1}{2}\pi)\;, I_{zb} = I_{z+}(-\xi, -\frac{1}{2}\pi), V_{\rho b} = V_{\rho +}(-\xi, -\frac{1}{2}\pi),\; I_{\rho b} = I_{\rho +}(-\xi, -\frac{1}{2}\pi) \tag{4.3.5.18}$$

Equations (4.3.5.11) and (4.3.5.12) can be rewritten as

$$\xi V_z - \frac{\tau_o^2}{\omega\varepsilon} I_\rho - \frac{\alpha_o \eta}{\omega\varepsilon} I_z = 0 \tag{4.3.5.19}$$

$$\xi I_z + \frac{\tau_o^2}{\omega\mu} V_\rho + \frac{\alpha_o \eta}{\omega\mu} V_z = 0 \tag{4.3.5.20}$$

where $V_z = V_{zp} + V_{za}$, $I_\rho = I_{\rho p} - I_{\rho a}$, $I_z = I_{zp} + I_{za}$, $V_\rho = V_{\rho p} - V_{\rho a}$ and thus (4.3.5.11) and (4.3.5.12) have the form

$$\begin{vmatrix} -V_\rho(\eta) \\ V_z(\eta) \end{vmatrix} = \mathrm{Z_c}(\eta) \cdot \begin{vmatrix} I_z(\eta) \\ I_\rho(\eta) \end{vmatrix} \tag{4.3.5.21}$$

where:

$$\mathbf{Z_c}(\eta) = \frac{1}{\omega\varepsilon\,\xi} \begin{vmatrix} k^2 - \eta^2 & \eta\,\alpha_o \\ \eta\,\alpha_o & \tau_o^2 \end{vmatrix} \tag{4.3.5.22}$$

The boundary conditions on face a impose

$$\begin{vmatrix} V_{\rho a}(\eta) \\ V_{za}(\eta) \end{vmatrix} = -Z_a \cdot \begin{vmatrix} I_{za}(\eta) \\ -I_{\rho a}(\eta) \end{vmatrix} \tag{4.3.5.23}$$

which yields

$$V_p = Z_c I_p + (Z_c + Z_a) I_{am} \tag{4.3.5.24}$$

where:

$$V_p = \begin{vmatrix} -V_{\rho p} \\ V_{zp} \end{vmatrix}, \quad I_p = \begin{vmatrix} I_{zp} \\ I_{\rho p} \end{vmatrix}, \quad I_{am} = \begin{vmatrix} I_{za} \\ -I_{\rho a} \end{vmatrix}, \tag{4.3.5.25}$$

We can obtain another equation by substituting $\eta \to -\eta$ and taking into account that

$$Z_c(-\eta) = \sigma\, Z_c(\eta)\, \sigma \tag{4.3.5.26}$$

with

$$\sigma = \begin{vmatrix} 1 & 0 \\ 0 & -1 \end{vmatrix} \tag{4.3.5.27}$$

It yields the independent equation

$$V_m = \sigma \cdot Z_c \cdot \sigma\, I_m + (\sigma \cdot Z_c \cdot \sigma + Z_a) I_{amp} \tag{4.3.5.28}$$

where:

$$V_m = \begin{vmatrix} -V_{\rho +}(-\eta, 0) \\ V_{z+}(-\eta, 0) \end{vmatrix}, \qquad I_m = \begin{vmatrix} I_{z+}(-\eta, 0) \\ I_{\rho +}(-\eta, 0) \end{vmatrix}, \quad I_{amp} = \begin{vmatrix} I_{z+}(\eta, \pi) \\ -I_{\rho +}(\eta, \pi) \end{vmatrix} \tag{4.3.5.29}$$

Multiplying (4.3.5.28) by σ and taking into account that

$$\sigma Z_a = Z_a \sigma \tag{4.3.5.30}$$

we have the compact form

$$\sigma \cdot V_m = Z_c \cdot \sigma\, I_m + (Z_c + Z_a)\sigma \cdot I_{amp} \tag{4.3.5.31}$$

Next, considering the PEC face b, from (4.3.5.13) and (4.3.5.14), we have

$$\xi V_{zp} + \frac{\tau_o^2}{\omega\varepsilon} I_{\rho p} + \frac{\alpha_o \eta}{\omega\varepsilon} I_{zp} = \frac{-I_{zb}\alpha_o \xi + I_{\rho b}\tau_o^2}{\omega\varepsilon} \tag{4.3.5.32}$$

$$\xi I_{zp} - \frac{\tau_o^2}{\omega\mu} V_{\rho p} - \frac{\alpha_o \eta}{\omega\mu} V_{zp} = \eta I_{zb} \tag{4.3.5.33}$$

Replacing η with $-\eta$ in (4.3.5.32)–(4.3.5.33), since ξ, I_{zb} and $I_{\rho b}$ do not change

$$\xi\ V_{zp} + \frac{\tau_o^2}{\omega\varepsilon} I_{\rho p} + \frac{\alpha_o \eta}{\omega\varepsilon} I_{zp} = \xi\ V_{zm} + \frac{\tau_o^2}{\omega\varepsilon} I_{\rho m} - \frac{\alpha_o \eta}{\omega\varepsilon} I_{zm} \tag{4.3.5.34}$$

$$\xi I_{zp} - \frac{\tau_o^2}{\omega\mu} V_{\rho p} - \frac{\alpha_o \eta}{\omega\mu} V_{zp} = -\xi I_{zm} + \frac{\tau_o^2}{\omega\mu} V_{\rho m} - \frac{\alpha_o \eta}{\omega\mu} V_{zm} \tag{4.3.5.35}$$

where the following four minus functions are introduced $V_{zm} = V_{z+}(-\eta, 0)$, $I_{zm} = I_{z+}(-\eta, 0)$, $V_{\rho m} = V_{\rho +}(-\eta, 0)$, $I_{\rho m} = I_{\rho +}(-\eta, 0)$.

The algebraic manipulations of (4.3.5.34) and (4.3.5.35) yield

$$V_{\rho p} + V_{\rho m} = \frac{(I_{zm} + I_{zp})(k^2 - \eta^2) + (I_{\rho p} - I_{\rho m})\alpha_o \eta}{\omega\varepsilon\xi} \tag{4.3.5.36}$$

$$V_{zp} - V_{zm} = -\frac{(I_{zm} + I_{zp})\alpha_o \eta + (I_{\rho p} - I_{\rho m})\tau_o^2}{\omega\varepsilon\xi} \tag{4.3.5.37}$$

or, equivalently, in a more compact form

$$V_p + \sigma \cdot V_m = -Z_c I_p - Z_c\, \sigma \cdot I_m \tag{4.3.5.38}$$

Equations (4.3.5.24), (4.3.5.31) and (4.3.5.38) constitute a system of $3\times 2 = 6$ classical WH equations with the plus functions $V_p, I_p, \sigma I_{amp}$ and the minus functions $\sigma V_m, \sigma I_m$, I_{am} as unknowns. Taking into account that, in this system, Z_a is diagonal, by separating the plus function and the minus functions, we obtain

$$V_p - \tilde{Z}_c (Z_a I_p) = (\tilde{Z}_c + 1)(Z_a I_{am}) \tag{4.3.5.39}$$

$$V_p + \tilde{Z}_c(Z_a I_p) = -\sigma \cdot V_m - \tilde{Z}_c(Z_a \sigma \cdot I_m) \tag{4.3.5.40}$$

$$(\tilde{Z}_c + 1)(Z_a \sigma \cdot I_{amp}) = \sigma \cdot V_m - \tilde{Z}_c \cdot (Z_a \sigma I_m) \tag{4.3.5.41}$$

or, equivalently, in a compact form

$$s \begin{vmatrix} V_p \\ Z_a I_p \\ Z_a \sigma I_{amp} \end{vmatrix} = d \begin{vmatrix} \sigma V_m \\ Z_a \sigma I_{amp} \\ Z_a I_{am} \end{vmatrix} \tag{4.3.5.42}$$

where:

$$V_p = \begin{vmatrix} -V_{\rho p} \\ V_{zp} \end{vmatrix}, \quad I_p = \begin{vmatrix} I_{zp} \\ I_{\rho p} \end{vmatrix}, \quad I_{am} = \begin{vmatrix} I_{za} \\ -I_{\rho a} \end{vmatrix}, \quad V_m = \begin{vmatrix} -V_{\rho m} \\ V_{zm} \end{vmatrix}, \quad I_m = \begin{vmatrix} I_{zm} \\ I_{\rho m} \end{vmatrix} \tag{4.3.5.43}$$

and

$$s = \begin{vmatrix} 1 & -\tilde{Z}_c & 0 \\ 1 & \tilde{Z}_c & 0 \\ 0 & 0 & \tilde{Z}_c + 1 \end{vmatrix} \quad d = \begin{vmatrix} 0 & 0 & 1 + \tilde{Z}_c \\ -1 & -\tilde{Z}_c & 0 \\ 1 & -\tilde{Z}_c & 0 \end{vmatrix} \tag{4.3.5.44}$$

$$\tilde{\mathrm{Z}}_c = \mathrm{Z}_c \cdot \mathrm{Z}_a^{-1} = \frac{1}{\omega \varepsilon \xi} \begin{vmatrix} k^2 - \eta^2 & \eta\, \alpha_o \\ \eta\, \alpha_o & \tau_o^2 \end{vmatrix} \mathrm{Z}_a^{-1} \tag{4.3.5.45}$$

The matrices s and d are of order 6, which are defined in terms of sub-matrices of order 2 (e.g. 1 represents the unit matrix of order 2). It is important to note that the sub-matrices of order 2 depend only on the matrix $\tilde{\mathbf{Z}}_c$. Thus, we can manipulate these sub-matrices as done for scalars. The evaluation of $d^{-1}s$ allows the factorization of the matrix kernel

$$G = d^{-1}s$$

$$= \begin{vmatrix} -\frac{1}{2} & -\frac{\tilde{Z}_c}{2} & \frac{1+\tilde{Z}_c}{2} \\ -\frac{1}{2\tilde{Z}_c} & -\frac{1}{2} & -\frac{1+\tilde{Z}_c}{2\tilde{Z}_c} \\ 1+\tilde{Z}_c & -\frac{\tilde{Z}_c}{1+\tilde{Z}_c} & 0 \end{vmatrix} = \begin{vmatrix} -\frac{1}{2} & 0 & \frac{1}{2} \\ 0 & -\frac{1}{2} & -\frac{1}{2} \\ \frac{1}{1-\tilde{Z}_c^2} & \frac{\tilde{Z}_c^2}{1-\tilde{Z}_c^2} & 0 \end{vmatrix} + \frac{1}{\tilde{Z}_c} \begin{vmatrix} 0 & -\frac{\tilde{Z}_c^2}{2} & \frac{\tilde{Z}_c^2}{2} \\ -\frac{1}{2} & 0 & -\frac{1}{2} \\ -\frac{\tilde{Z}_c^2}{1-\tilde{Z}_c^2} & -\frac{\tilde{Z}_c^2}{1-\tilde{Z}_c^2} & 0 \end{vmatrix}$$

$$= R_o + \frac{1}{\tilde{Z}_c} R_1 \tag{4.3.5.46}$$

The rational matrix R_o has the Jordan form

$$R_o = \begin{vmatrix} -\frac{1}{2} & 0 & \frac{1}{2} \\ 0 & -\frac{1}{2} & -\frac{1}{2} \\ \frac{1}{1-\tilde{Z}_c^2} & \frac{\tilde{Z}_c^2}{1-\tilde{Z}_c^2} & 0 \end{vmatrix} = ABA^{-1} \tag{4.3.5.47}$$

where:

$$A = \begin{vmatrix} -1 & -\tilde{Z}_c^2 & \frac{1}{2} \\ 1 & 1 & -\frac{1}{2} \\ 1 & 0 & 1 \end{vmatrix}, \quad B = \begin{vmatrix} -1 & 0 & 0 \\ 0 & -\frac{1}{2} & 0 \\ 0 & 0 & \frac{1}{2} \end{vmatrix} \tag{4.3.5.48}$$

Since A is a polynomial matrix, the factorization of G reduces to that of

$$g = B + \frac{1}{\tilde{Z}_c} A^{-1} R_1 A = \tag{4.3.5.49}$$

where:

$$R_1 = \begin{vmatrix} 0 & -\frac{\tilde{Z}_c^2}{2} & \frac{\tilde{Z}_c^2}{2} \\ -\frac{1}{2} & 0 & -\frac{1}{2} \\ -\frac{\tilde{Z}_c^2}{1-\tilde{Z}_c^2} & -\frac{\tilde{Z}_c^2}{1-\tilde{Z}_c^2} & 0 \end{vmatrix} \tag{4.3.5.50}$$

After algebraic manipulations, we get

$$g = \begin{vmatrix} -1 & 0 & 0 \\ 0 & -\frac{1}{2} & -\frac{3}{4\tilde{Z}_c} \\ 0 & -\tilde{Z}_c & \frac{1}{2} \end{vmatrix} \tag{4.3.5.51}$$

The factorization of the matrix g of order 6 reduces to the factorization of the matrix of order 4:

$$m = \begin{vmatrix} -\frac{1}{2} & -\frac{3}{4\tilde{Z}_c} \\ -\tilde{Z}_c & \frac{1}{2} \end{vmatrix} \tag{4.3.5.52}$$

where we recall that $\tilde{Z}_c = Z_c Z_a^{-1}$ $\tilde{Z}_c^{-1} = Z_a Z_c^{-1}$.

The relationship

$$\begin{vmatrix} j\sqrt{2}\,Z_a^{-1} & 0 \\ 0 & \sqrt{2} \end{vmatrix} \begin{vmatrix} -\frac{1}{2} & -\frac{3}{4\tilde{Z}_c} \\ -\tilde{Z}_c & \frac{1}{2} \end{vmatrix} \begin{vmatrix} j\sqrt{2}\,Z_a & 0 \\ 0 & \sqrt{2} \end{vmatrix} = \begin{vmatrix} 1 & -j\frac{3}{2} Z_c^{-1} \\ -j2\,Z_c & 1 \end{vmatrix} \tag{4.3.5.53}$$

reduces m to the factorization of the matrix of order 4:

$$\begin{vmatrix} 1 & -j\frac{3}{2} Y_c \\ -j2\,Z_c & 1 \end{vmatrix} \tag{4.3.5.54}$$

The key point in reducing the order of the matrices to be factorized is the observation that the polynomial matrix $t(\eta)=\begin{vmatrix}\eta & -\alpha_o\\ \alpha_o & \eta\end{vmatrix}$ makes the matrices Z_c and Y_c diagonal. In fact

$$Z_c = t^{-1}(\eta)\cdot\begin{vmatrix}\frac{\xi}{\omega\varepsilon} & 0\\ 0 & \frac{k^2}{\omega\varepsilon\xi}\end{vmatrix}\cdot t(\eta),\qquad \mathbf{Y}_c = t^{-1}(\eta)\begin{vmatrix}\frac{\omega\varepsilon}{\xi} & 0\\ 0 & \frac{\xi}{\omega\mu}\end{vmatrix}t(\eta) \tag{4.3.5.55}$$

Consequently, we reduce the factorization of

$$\begin{vmatrix}1 & -j\frac{3}{2}Y_c\\ -j2Z_c & 1\end{vmatrix} \tag{4.3.5.56}$$

to that of

$$M=\begin{vmatrix}1 & -j\frac{3}{2}y_c\\ -j2z_c & 1\end{vmatrix} \tag{4.3.5.57}$$

where $z_c(\eta)=\begin{vmatrix}\frac{\xi}{\omega\varepsilon} & 0\\ 0 & \frac{\omega\mu}{\xi}\end{vmatrix}$, $y_c(\eta)=\begin{vmatrix}\frac{\omega\varepsilon}{\xi} & 0\\ 0 & \frac{\xi}{\omega\mu}\end{vmatrix}$. By rewriting in terms of scalar elements, we have

$$M=\begin{vmatrix}1 & 0 & -j\frac{3}{2}\frac{\xi}{\omega\varepsilon} & 0\\ 0 & 1 & 0 & -j\frac{3}{2}\frac{\omega\mu}{\xi}\\ -j2\frac{\omega\varepsilon}{\xi} & 0 & 1 & 0\\ 0 & -j2\frac{\xi}{\omega\mu} & 0 & 1\end{vmatrix} \tag{4.3.5.58}$$

Rows 1 and 3 are decoupled from rows 2 and 4. It means that M reduces to the factorization of two matrices of order 2:

$$\begin{vmatrix} 1 & -j\frac{3}{2}\frac{\xi}{\omega\varepsilon} \\ -j2\frac{\omega\varepsilon}{\xi} & 1 \end{vmatrix}, \text{ and } \begin{vmatrix} 1 & -j\frac{3}{2}\frac{\omega\mu}{\xi} \\ -j2\frac{\xi}{\omega\mu} & 1 \end{vmatrix} \tag{4.3.5.59}$$

The two matrices reported in (4.3.5.59) are of the Daniele–Khrapkov type, thus their factorizations become straightforward.

4.4. Exact solutions for impedance wedge problems with the GWHE form of section 3.5 – form #1

Before discussing particular values of the impedance faces that allow explicit solutions of impedance wedge problems, we again note that explicit solutions are often very difficult to obtain. According to the authors' opinion, more practical solutions can be obtained by using the general Fredholm factorization method, as described and applied in Chapter 5. However, new exact solutions are always welcome to extend the mathematical methodology in the WH technique and to validate the efficiency of approximate solution methods.

The WH solutions of impedance wedge problems are at least the same as those obtained using the SM method. In order to demonstrate this claim, in sections 4.4 and 4.5, we will focus our exposition on the WH solutions of problems which have been already solved in the literature using the SM method.

Using the reference system of Figure 4.4.1, the mapping (3.4.1) reduces the GWHEs of a generic impedance wedge to a classical system of WH equations in the $\bar{\eta}$ plane (see Chapter 3).

For a wedge of an arbitrary aperture angle, we note that the complexity of the solution of the scattering problem is higher at skew incidence ($\alpha_o \neq 0$) than at normal incidence $\alpha_o = 0$.

In particular, in section 4.4, we will use a mathematical form of the GWHEs (see section 3.5) that is slightly different from the one presented in sections 4.1 and 4.5. The three different forms of GWHEs have advantages for the exact solution of particular impenetrable wedge problems. In particular, we note that the form used in section 4.4 is the one to which we apply the Fredholm factorization method.

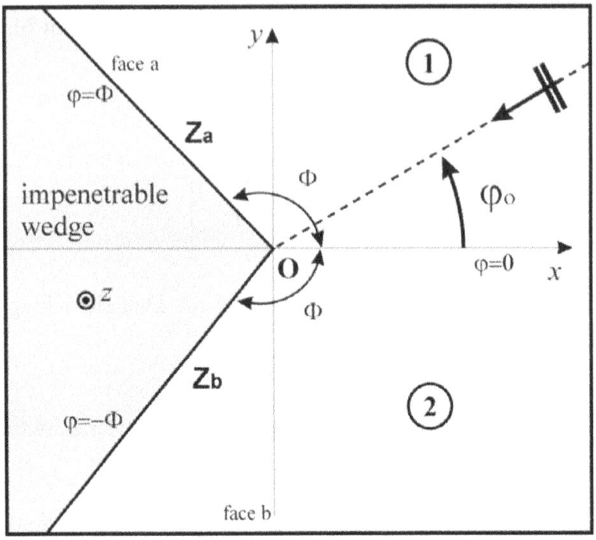

Figure 4.4.1. *Impenetrable wedge with arbitrary aperture and face impendences at* $\varphi = \pm\Phi$. *The problem is described in terms of Cartesian coordinates along with cylindrical coordinates, and we assume the wedge with cylindrical symmetry with respect to the z axis*

4.4.1. *The WH solution of the Malyuzhinets problem*

The Malyuzhinets problem represents a fundamental problem in the theory of diffraction by impedance wedges. In fact, to solve this problem, Malyuzhinets introduced a very ingenious theory (the SM technique) (Malyuzhinets 1958), which has produced very important advances in the theory of diffraction. The impedance wedge considered by Malyuzhinets is the particular one where the two face impedances are isotropic $\mathbf{Z}^a = Z_o z_a \mathbf{1}$, $\mathbf{Z}^b = Z_o z_b \mathbf{1}$, although different, i.e. the isotropic non-symmetric impedance case. Furthermore, the incidence of the impinging plane wave is the normal one: $\alpha_o = 0$. The study by Daniele (2003b), which is the fundamental base of this book, was dedicated, in particular, to the solution of the Malyuzhinets problem using the WH technique.

The general equations (3.5.8) given in section 3.5 of Chapter 3 simplify at normal incidence $(\alpha_o = 0, \tau_o = k)$ with isotropic impedances, yielding the system of GWHEs in the matrix form

$$G(\eta)X_+(\eta) = X_-(m) \qquad (4.4.1.1)$$

$$X_+(\eta)=\begin{vmatrix} V_{z+}(\eta,0) \\ V_{\rho+}(\eta,0) \\ Z_o I_{z+}(\eta,0) \\ Z_o I_{\rho+}(\eta,0) \end{vmatrix}, \qquad X_-(m)=\begin{vmatrix} Z_o I_{\rho+}(-m,\Phi) \\ -Z_o I_{z+}(-m,\Phi) \\ -Z_o I_{\rho+}(-m,-\Phi) \\ Z_o I_{z+}(-m,-\Phi) \end{vmatrix} \tag{4.4.1.2}$$

$$G(\eta)=D^{-1}(\eta)S(\eta)=\begin{vmatrix} -\dfrac{\xi}{k+n z_a} & 0 & 0 & \dfrac{k}{k+n z_a} \\ 0 & \dfrac{k}{kz_a+n} & \dfrac{\xi}{kz_a+n} & 0 \\ -\dfrac{\xi}{k+n z_b} & 0 & 0 & -\dfrac{k}{k+n z_b} \\ 0 & \dfrac{k}{kz_b+n} & -\dfrac{\xi}{kz_b+n} & 0 \end{vmatrix} \tag{4.4.1.3}$$

with $m(\eta)=-\eta\cos\Phi+\xi(\eta)\sin\Phi$ and $n=-\eta\sin\Phi-\xi\cos\Phi$.

We decouple the GWHEs of order 4 into two independent systems of order 2, by taking into account that

$$\begin{vmatrix} 1 & 0 & 0 & 0 \\ 0 & 0 & 1 & 0 \\ 0 & 0 & 0 & 1 \\ 0 & 1 & 0 & 0 \end{vmatrix} G(\eta) \begin{vmatrix} 1 & 0 & 0 & 0 \\ 0 & 0 & 1 & 0 \\ 0 & 0 & 0 & 1 \\ 0 & 1 & 0 & 0 \end{vmatrix} = \begin{vmatrix} -\dfrac{\xi}{k+n z_a} & \dfrac{k}{k+n z_a} & 0 & 0 \\ -\dfrac{\xi}{k+n z_b} & -\dfrac{k}{k+n z_b} & 0 & 0 \\ 0 & 0 & \dfrac{k}{n+k z_b} & -\dfrac{\xi}{n+k z_b} \\ 0 & 0 & \dfrac{k}{n+k z_a} & \dfrac{\xi}{n+k z_a} \end{vmatrix} \tag{4.4.1.4}$$

Next, for simplicity, we consider only the first system of order 2 that corresponds to the case of Ez-polarization:

$$G_\Phi(\eta)F_+(\eta)=F_-(m) \tag{4.4.1.5}$$

where:

$$F_+(\eta)=\begin{vmatrix} V_{z+}(\eta,0) \\ Z_o I_{\rho+}(\eta,0) \end{vmatrix}, \qquad F_-(m)=\begin{vmatrix} Z_o I_{\rho+}(-m,\Phi) \\ -Z_o I_{\rho+}(-m,-\Phi) \end{vmatrix} \tag{4.4.1.6}$$

Taking into account section 3.6 of Chapter 3, the solution of the GWHEs reduces to a classical factorization of the matrix kernel $\bar{G}_\Phi(\bar{\eta}) = G_\Phi(\eta)$ in the $\bar{\eta}$ plane (3.4.1):

$$\bar{G}_\Phi(\bar{\eta}) = \begin{vmatrix} -\dfrac{\bar{\xi}}{k+\bar{n}\, z_a} & \dfrac{k}{k+\bar{n}\, z_a} \\ -\dfrac{\bar{\xi}}{k+\bar{n}\, z_b} & -\dfrac{k}{k+\bar{n}\, z_b} \end{vmatrix} \tag{4.4.1.7}$$

We obtain an explicit factorization by reducing $\bar{G}_\Phi(\bar{\eta})$ to a matrix of the Daniele–Khrapkov type. For this task, algebraic manipulations allow us to rewrite $\bar{G}_\Phi(\bar{\eta})$ in the form:

$$\bar{G}_\Phi(\bar{\eta}) = \frac{2\bar{n}+n_a+n_b}{2(\bar{n}+n_a)(\bar{n}+n_b)} \begin{bmatrix} y_a & y_a \\ y_b & -y_b \end{bmatrix} \begin{bmatrix} \bar{\xi}_- & 0 \\ 0 & \sqrt{k} \end{bmatrix} \begin{bmatrix} 1 & -b \\ -a & 1 \end{bmatrix} \begin{bmatrix} \bar{\xi}_+ & 0 \\ 0 & \sqrt{k} \end{bmatrix} \begin{bmatrix} -1 & 0 \\ 0 & 1 \end{bmatrix} \tag{4.4.1.8}$$

where:

$$b = f(\eta)\frac{\sqrt{k}}{\bar{\xi}_-(\bar{\eta})}, \qquad a = f(\eta)\frac{\bar{\xi}_-(\bar{\eta})}{\sqrt{k}}, \qquad f(\eta) = \frac{n_a - n_b}{2\bar{n}+n_a+n_b},$$

$$n_{a,b} = \frac{\omega\mu}{Z_o z_{a,b}} = k y_{a,b} = k\sin\vartheta_{a,b} \tag{4.4.1.9}$$

Taking into account that the ratio

$$-b/(-a) = 2k/(k+\bar{\eta}) = N(\bar{\eta})/P(\bar{\eta}) \tag{4.4.1.10}$$

is rational, we introduce the polynomials $N(\bar{\eta}) = 2k$ and $P(\bar{\eta}) = k+\bar{\eta}$ and, to factorize (4.4.1.8), we use the identity (see section 1.8, Chapter 1):

$$\begin{bmatrix} 1 & -b \\ -a & 1 \end{bmatrix} = \sqrt{1-ab} \begin{vmatrix} \cosh[\frac{1}{2}t\,(\bar{\eta})\sqrt{N(\bar{\eta})P(\bar{\eta})}] & \sqrt{\dfrac{N(\bar{\eta})}{P(\bar{\eta})}}\sinh[\frac{1}{2}t\,(\bar{\eta})\sqrt{N(\bar{\eta})P(\bar{\eta})}] \\ \sqrt{\dfrac{P(\bar{\eta})}{N(\bar{\eta})}}\sinh[\frac{1}{2}t\,(\bar{\eta})\sqrt{N(\bar{\eta})P(\bar{\eta})}] & \cosh[\frac{1}{2}t\,(\bar{\eta})\sqrt{N(\bar{\eta})P(\bar{\eta})}] \end{vmatrix} \tag{4.4.1.11}$$

where:

$$\sqrt{1-ab}=\left[\frac{2\overline{n}+n_a+n_b}{2\sqrt{(\overline{n}+n_a)(\overline{n}+n_b)}}\right]^{-1} \tag{4.4.1.12}$$

$$t\,(\overline{\eta})k=\frac{1}{\sqrt{N(\overline{\eta})P(\overline{\eta})}}\log\frac{1+\sqrt{a(\overline{\eta})b(\overline{\eta})}}{1+\sqrt{a(\overline{\eta})b(\overline{\eta})}}=\frac{1}{\sqrt{2k(k+\overline{\eta})}}\log\frac{\overline{n}+n_b}{\overline{n}+n_a}=\frac{1}{2\sqrt{k}\overline{\xi}_-}\log\frac{\overline{n}+n_b}{\overline{n}+n_a} \tag{4.4.1.13}$$

From (4.4.1.8), we must factorize the scalars $\overline{\xi}(\overline{\eta})$, $\overline{n}$ and

$$\overline{g}(\overline{\eta})=\frac{2\overline{n}+n_a+n_b}{2(\overline{n}+n_a)(\overline{n}+n_b)}\sqrt{1-ab}=\frac{1}{\sqrt{(\overline{n}+n_a)(\overline{n}+n_b)}} \tag{4.4.1.14}$$

Furthermore, we must decompose the scalar $t\,(\overline{\eta})$ (4.4.1.13).

The factorization of $\overline{\xi}$ and $\overline{n}$ is presented in Daniele and Zich (2014) and in sections 3.8 and 6.3:

$$\xi_-(\overline{\eta})=\sqrt{\frac{k+\overline{\eta}}{2}},\qquad \overline{\xi}_+(\overline{\eta})=\frac{\xi(\eta(\overline{\eta}))}{\overline{\xi}_-(\overline{\eta})} \tag{4.4.1.15}$$

$$\overline{n}_+(\overline{\eta})=\sqrt{\frac{k-\overline{\eta}}{2}},\qquad \overline{n}_-(\overline{\eta})=\frac{n(\eta(\overline{\eta}))}{\overline{n}_+(\overline{\eta})} \tag{4.4.1.16}$$

To obtain the factorization of $\overline{g}(\overline{\eta})$, we rewrite it in the form:

$$\overline{g}(\overline{\eta})=\frac{1}{\overline{n}\sqrt{\frac{(\overline{n}+n_a)}{n}}\sqrt{\frac{(\overline{n}+n_b)}{n}}} \tag{4.4.1.17}$$

Using the procedure described in Appendix 4.A, we obtain:

$$\left[\frac{\overline{n}+n_{a,b}}{\overline{n}}\right]_+=d_\Phi(\overline{\eta})=\exp\left[\frac{1}{\pi}\int_0^\infty \arctan[\frac{\sin\vartheta_{a,b}}{\sinh[\frac{\Phi}{\pi}u]}]\frac{\sinh u}{\cosh u-\overline{\eta}/k}du\right] \tag{4.4.1.18}$$

where $\sin\vartheta_{a,b} = 1/z_{a,b} = y_{a,b}$. Using the same procedure, we also obtain the decomposition of

$$t(\bar{\eta}) = \frac{1}{2\sqrt{k}}\bar{t}(\bar{\eta}) \tag{4.4.1.19}$$

We have:

$$\begin{aligned}\bar{t}_+(\bar{\eta}) &= \left\{\frac{1}{\bar{\xi}_-(\bar{\eta})}\log[\frac{n+n_b}{n+n_a}]\right\}_+ \\ &= -\frac{2}{\pi}\int_0^\infty \left[\arctan[\frac{\sin\vartheta_b}{\sinh[\frac{\Phi}{\pi}u]}] - \arctan[\frac{\sin\vartheta_a}{\sinh[\frac{\Phi}{\pi}u]}]\right]\frac{\sinh\frac{u}{2}}{\cosh u - \bar{\eta}/k}du\end{aligned} \tag{4.4.1.20}$$

From the above scalar factorization and decomposition, we obtain the factorized matrices of (4.4.1.8) in the form:

$$\bar{G}_{\Phi-}(\bar{\eta}) = \bar{g}_-(\bar{\eta})\begin{vmatrix} \cosh[\frac{1}{2}t_-(\bar{\eta})\sqrt{N(\bar{\eta})P(\bar{\eta})}] & \sqrt{\frac{N(\bar{\eta})}{P(\bar{\eta})}}\sinh[\frac{1}{2}t_-(\bar{\eta})\sqrt{N(\bar{\eta})P(\bar{\eta})}] \\ \sqrt{\frac{P(\bar{\eta})}{N(\bar{\eta})}}\sinh[\frac{1}{2}t_-(\bar{\eta})\sqrt{N(\bar{\eta})P(\bar{\eta})}] & \cosh[\frac{1}{2}t_-(\bar{\eta})\sqrt{N(\bar{\eta})P(\bar{\eta})}] \end{vmatrix}\begin{bmatrix} y_a & y_a \\ y_b & -y_b \end{bmatrix}\begin{bmatrix} \bar{\xi}_- & 0 \\ 0 & \sqrt{k} \end{bmatrix} \tag{4.4.1.21}$$

$$\bar{G}_{\Phi+}(\bar{\eta}) = \bar{g}_+(\bar{\eta})\begin{vmatrix} \cosh[\frac{1}{2}t_+(\bar{\eta})\sqrt{N(\bar{\eta})P(\bar{\eta})}] & \sqrt{\frac{N(\bar{\eta})}{P(\bar{\eta})}}\sinh[\frac{1}{2}t_+(\bar{\eta})\sqrt{N(\bar{\eta})P(\bar{\eta})}] \\ \sqrt{\frac{P(\bar{\eta})}{N(\bar{\eta})}}\sinh[\frac{1}{2}t_+(\bar{\eta})\sqrt{N(\bar{\eta})P(\bar{\eta})}] & \cosh[\frac{1}{2}t_+(\bar{\eta})\sqrt{N(\bar{\eta})P(\bar{\eta})}] \end{vmatrix}\begin{bmatrix} \bar{\xi}_+ & 0 \\ 0 & \sqrt{k} \end{bmatrix}\begin{bmatrix} -1 & 0 \\ 0 & 1 \end{bmatrix} \tag{4.4.1.22}$$

Note that $\bar{g}_-(\bar{\eta})$ and $\bar{g}_+(\bar{\eta})$ behave as constants for $\bar{\eta} \to \infty$. In addition, taking into account that $\eta \approx \bar{\eta}^{\Phi/\pi}$, it follows that $\bar{t}_\pm(\bar{\eta}) = O[\frac{1}{\bar{\eta}^{1/2}\bar{\eta}^{\Phi/\pi}}]$. Since the argument of the hyperbolic functions in (4.4.1.21) and (4.4.1.22) is constant or vanishing as $\bar{\eta} \to \infty$, both the factorized matrices are bounded at $\bar{\eta} \to \infty$. Finally, the structure of the second members avoids the offensive presence of the singularities of $\sqrt{N(\bar{\eta})P(\bar{\eta})}$ and $\sqrt{\frac{N(\bar{\eta})}{P(\bar{\eta})}}$ or $\sqrt{\frac{P(\bar{\eta})}{N(\bar{\eta})}}$.

To complete the solution process using (1.5.13), we need to evaluate the non-standard contributions of the geometrical optical field. Considering the pole $\bar{\eta}_o = -k\cos\frac{\pi}{\Phi}\varphi_o$ as non-standard for the minus function $\bar{F}_-(\bar{\eta}) = F_-(m(\eta(\bar{\eta})))$, we have

$$\bar{F}_-^{n.s.}(\bar{\eta}) = \bar{F}_-^{G.O.}(\bar{\eta}) = \left(\bar{G}_\Phi(\bar{\eta}_o)\right)^{-1}\frac{R_o}{\bar{\eta}-\bar{\eta}_o} \tag{4.4.1.23}$$

where:

$$R_o = \left| \begin{matrix} -\dfrac{\pi}{\Phi}\dfrac{2j\sin\frac{\pi}{\Phi}\varphi_o}{Z_o + Z_a\sin(\Phi-\varphi_o)}E_o \\ 0 \end{matrix} \right| \tag{4.4.1.24}$$

For a discussion on the non-standard part resulting from GO fields, see section 3.7.

From (1.5.13), the solution is

$$\bar{F}_+(\bar{\eta}) = \left(\bar{G}_{\Phi+}(\bar{\eta})\right)^{-1}\bar{G}_{\Phi+}(\bar{\eta}_o)\left(\bar{G}_\Phi(\bar{\eta}_o)\right)^{-1}\frac{R_o}{\bar{\eta}-\bar{\eta}_o} \tag{4.4.1.25}$$

Although the numerical evaluation of (4.4.1.18) and (4.4.1.20) can be performed without difficulty, it is worth noting that they are related to the Malyuzhinets function $\Psi_\Phi(w)$ (Appendix 4.A). In fact, cumbersome mathematical deductions (which are not reported here) yield two relations of the factorization of $n+n_{a,b}$ and the decomposition of $t(\bar{\eta})$ with the function $\Psi_\Phi(w)$:

$$\left[\frac{n+n_{a,b}}{\sqrt{k}}\right]_+\Bigg|_{\eta=-k\cos w}$$
$$= d_\Phi(0)\frac{\Psi_\Phi(w+\Phi+\tilde{\vartheta}_{a,b})\Psi_\Phi(w+\Phi-\tilde{\vartheta}_{a,b})\Psi_\Phi(w-\Phi+\tilde{\vartheta}_{a,b})\Psi_\Phi(w-\Phi-\tilde{\vartheta}_{a,b})}{\Psi_\Phi(\Phi+\tilde{\vartheta}_{a,b})\Psi_\Phi(\Phi-\tilde{\vartheta}_{a,b})\Psi_\Phi(-\Phi+\tilde{\vartheta}_{a,b})\Psi_\Phi(-\Phi-\tilde{\vartheta}_{a,b})} \tag{4.4.1.26}$$

$$\left\{\frac{\sqrt{k}}{\xi_-(\overline{\eta})}\log[\frac{n+n_b}{n+n_a}]\right\}_+\Bigg|_{\eta=-k\cos w}$$

$$=\frac{1}{\sin[\frac{\pi\, w}{2\Phi}]}\log\left[\frac{\Psi_\Phi(w-\Phi-\tilde{\vartheta}_b)\Psi_\Phi(w-\Phi+\tilde{\vartheta}_b)\Psi_\Phi(w+\Phi-\tilde{\vartheta}_a)\Psi_\Phi(w+\Phi+\tilde{\vartheta}_a)}{\Psi_\Phi(w+\Phi-\tilde{\vartheta}_b)\Psi_\Phi(w+\Phi+\tilde{\vartheta}_b)\Psi_\Phi(w-\Phi-\tilde{\vartheta}_a)\Psi_\Phi(w-\Phi+\tilde{\vartheta}_a)}\right] \tag{4.4.1.27}$$

with $\eta=-k\cos w$, $w=\frac{\Phi}{\pi}\overline{w}$, $\tilde{\vartheta}_{a,b}=\frac{\pi}{2}-\vartheta_{a,b}$. These alternative expressions are important, since they allow us to demonstrate the equivalence of the Wiener-Hopf solution (4.4.1.25) with the one obtained using the SM method (Malyuzhinets 1958).

Numerical examples of this problem are presented in sections 5.9 and 5.10 of Chapter 5, comparing the WH solution given in this section with those obtained using the Fredholm factorization method.

4.4.2. Diffraction at skew incidence ($\alpha_o \neq 0$) by a wedge with a PEC and a PMC face

From Figure 4.1.1, by applying the PEC boundary condition at face a and the PMC boundary condition at face b, functional equations (3.3.57)–(3.3.60) for angular regions 1 and 2 become:

$$\xi V_{z+}(\eta,0)-\frac{\tau_o^2}{\omega\varepsilon}I_{\rho+}(\eta,0)-\frac{\alpha_o\eta}{\omega\varepsilon}I_{z+}(\eta,0)=-\frac{\tau_o^2}{\omega\varepsilon}I_{\rho+}(-m,\Phi)+\frac{\alpha_o m}{\omega\varepsilon}I_{z+}(-m,\Phi) \tag{4.4.2.1}$$

$$\xi I_{z+}(\eta,0)+\frac{\tau_o^2}{\omega\mu}V_{\rho+}(\eta,0)+\frac{\alpha_o\eta}{\omega\mu}V_{z+}(\eta,0)=-nI_{z+}(-m,\Phi) \tag{4.4.2.2}$$

$$\xi V_{z+}(\eta,0)+\frac{\tau_o^2}{\omega\varepsilon}I_{\rho+}(\eta,0)+\frac{\alpha_o\eta}{\omega\varepsilon}I_{z+}(\eta,0)=-nV_{z+}(-m,-\Phi) \tag{4.4.2.3}$$

$$\xi I_{z+}(\eta,0)-\frac{\tau_o^2}{\omega\mu}V_{\rho+}(\eta,0)-\frac{\alpha_o\eta}{\omega\mu}V_{z+}(\eta,0)=-\frac{\tau_o^2}{\omega\mu}V_{\rho+}(-m,-\Phi)+\frac{\alpha_o m}{\omega\mu}V_{z+}(-m,-\Phi) \tag{4.4.2.4}$$

These GWHEs can be rewritten in the form:

$$\begin{vmatrix} \xi 1 & 1 \\ \xi 1 & -1 \end{vmatrix} X_+(\eta) = \begin{vmatrix} d_1 & 0 \\ 0 & d_2 \end{vmatrix} Y_+(-m) \tag{4.4.2.5}$$

where:

$$X_+(\eta) = \begin{vmatrix} V_{z+}(\eta,0) \\ I_{z+}(\eta,0) \\ -\dfrac{\tau_o^2}{\omega\varepsilon} I_{\rho+}(\eta,0) - \dfrac{\alpha_o \eta}{\omega\varepsilon} I_{z+}(\eta,0) \\ \dfrac{\tau_o^2}{\omega\mu} V_{\rho+}(-\eta,0) + \dfrac{\alpha_o \eta}{\omega\mu} V_{z+}(\eta,0) \end{vmatrix},$$

$$Y_+(-m) = \begin{vmatrix} -\dfrac{\tau_o}{\omega\varepsilon} I_{\rho+}(-m,\Phi) + \dfrac{\alpha_o m}{\tau_o \omega\varepsilon} I_{z+}(-m,\Phi) \\ -I_{z+}(-m,\Phi) \\ -V_{z+}(-m,\Phi) \\ -\dfrac{\tau_o}{\omega\mu} V_{\rho+}(-m,-\Phi) + \dfrac{\alpha_o m}{\tau_o \omega\mu} V_{z+}(-m,-\Phi) \end{vmatrix} \tag{4.4.2.6}$$

with $d_1 = \begin{vmatrix} \tau_o & 0 \\ 0 & n \end{vmatrix}$ and $d_2 = \begin{vmatrix} n & 0 \\ 0 & \tau_o \end{vmatrix}$ being invertible. Applying (3.4.1) to (4.4.2.5), we have

$$\begin{vmatrix} \bar{\xi} \mathbf{1} & \mathbf{1} \\ \bar{\xi} \mathbf{1} & -\mathbf{1} \end{vmatrix} \bar{X}_+(\bar{\eta}) = \begin{vmatrix} \bar{\mathbf{d}}_1 & 0 \\ 0 & \bar{\mathbf{d}}_2 \end{vmatrix} \bar{Y}_-(\bar{\eta}) \tag{4.4.2.7}$$

The factorization of the matrix kernel of (4.4.2.7) involves a matrix commutative algebra, which is studied in section 1.8. It yields the form

$$\begin{vmatrix} \bar{\xi}_- \bar{n}_- & 0 \\ 0 & n_- \end{vmatrix} \begin{vmatrix} 1 & \dfrac{1}{\bar{\xi}_-} \bar{f}(\bar{\eta}) \\ \bar{\xi}_- \bar{f}(\bar{\eta}) & 1 \end{vmatrix} \begin{vmatrix} \bar{\xi}_+ \bar{n}_+ & 0 \\ 0 & \bar{n}_+ \end{vmatrix} \tag{4.4.2.8}$$

where $\bar{D}_\Sigma = \dfrac{\bar{n} + \tau_o}{2\bar{n}\tau_o} \begin{vmatrix} 1 & 0 \\ 0 & 1 \end{vmatrix}$, $\bar{D}_\Delta = \dfrac{\bar{n} - \tau_o}{2\bar{n}\tau_o} \begin{vmatrix} 1 & 0 \\ 0 & -1 \end{vmatrix}$, $\bar{f}(\bar{\eta}) = \dfrac{\bar{n} - \tau_o}{\bar{n} + \tau_o} \begin{vmatrix} 1 & 0 \\ 0 & -1 \end{vmatrix}$.

In particular, the central matrix of (4.4.2.8) G_d has the form (1.8.19), where $\bar{a}/\bar{b} = \frac{1}{\bar{\xi}_-^2} = \frac{2}{\tau_o + \bar{\eta}}$.

The explicit factorization is presented in section 1.8, yielding

$$G_{d\mp} = \begin{vmatrix} \left[\sqrt{1-\bar{f}^2}\right]_\mp & 0 \\ 0 & \left[\sqrt{1-\bar{f}^2}\right]_\mp \end{vmatrix} \begin{vmatrix} \cosh\left[\frac{1}{2}\bar{\xi}_-\bar{t}_\mp\right] & \frac{1}{\bar{\xi}_-}\sinh\left[\frac{1}{2}\bar{\xi}_-\bar{t}_\mp\right] \\ \bar{\xi}_-\cosh\left[\frac{1}{2}\bar{\xi}_-\bar{t}_\mp\right] & \cosh\left[\frac{1}{2}\bar{\xi}_-\bar{t}_\mp\right] \end{vmatrix} \quad (4.4.2.9)$$

where $\bar{t}_\mp = \left\{\frac{1}{\bar{\xi}_-}\log\frac{1+\bar{f}}{1-\bar{f}}\right\}_\mp$.

The factorization of the diagonal matrix $\sqrt{1-\bar{f}^2}$ and the decomposition of the diagonal matrix $t = \frac{1}{\bar{\xi}_-}\log\frac{1+\bar{f}}{1-\bar{f}}$ must be accomplished in the $\bar{\eta}$ plane. Since f is defined in the m plane, we must apply the substitutions

$$m = \tau_o \cos(w + \Phi) = \tau_o \cos(\frac{\Phi}{\pi}\bar{w} + \Phi) = \tau_o \cos[\frac{\Phi}{\pi}\arccos(-\frac{\bar{\eta}}{\tau_o}) + \Phi] \quad (4.4.2.10)$$

to obtain $\bar{f}(\bar{\eta})$. The process of factorization and decomposition in the $\bar{\eta}$ plane can be simplified by introducing the $\bar{w}$ plane defined by $\bar{\eta} = -\tau_o \cos\bar{w}$ (see Chapter 3 and Appendix 4.A). Thus, it is convenient to consider $\tilde{f}(\bar{w}) = f(\tau_o \cos(\frac{\Phi}{\pi}\bar{w} + \Phi))$. We also note that the factorized matrices present algebraic behavior for $\bar{\eta} \to \infty$. This means that, in the framework of the WH technique, we deal with a simple problem.

4.4.3. *Diffraction at skew incidence ($\alpha_o \neq 0$) by a wedge with a PEC face and the other face with diagonal Z^b with one null element*

This problem was solved using the SM method in Manara *et al.* (2000). In particular, from Figure 4.4.1, we assume a PEC impedance boundary condition for the face a and a diagonal Z^b with one null element as the boundary condition for the face b.

Taking into account that

$$Z^a = 0, \qquad Z^b = \begin{vmatrix} 0 & 0 \\ 0 & Z_f \end{vmatrix} \tag{4.4.3.1}$$

functional equations (3.3.57)–(3.3.60) become:

$$\xi V_{z+}(\eta,0) - \frac{\tau^2}{\omega\varepsilon} I_{\rho+}(\eta,0) - \frac{\alpha_o \eta}{\omega\varepsilon} I_{z+}(\eta,0) = -\frac{\tau^2}{\omega\varepsilon} I_{\rho+}(-m,\Phi) + \frac{\alpha_o m}{\omega\varepsilon} I_{z+}(-m,\Phi) \tag{4.4.3.2}$$

$$\xi I_{z+}(\eta,0) + \frac{\tau_o^2}{\omega\mu} V_{\rho+}(\eta,0) + \frac{\alpha_o \eta}{\omega\mu} V_{z+}(\eta,0) = -nI_{z+}(-m,\Phi) \tag{4.4.3.3}$$

$$\xi V_{z+}(\eta,0) + \frac{\tau_o^2}{\omega\varepsilon} I_{\rho+}(\eta,0) + \frac{\alpha_o \eta}{\omega\varepsilon} I_{z+}(\eta,0) = \frac{\tau_o^2}{\omega\varepsilon} I_{\rho+}(-m,-\Phi) - \frac{\alpha_o m}{\omega\varepsilon} I_{z+}(-m,-\Phi) \tag{4.4.3.4}$$

$$\xi I_{z+}(\eta,0) - \frac{\tau_o^2}{\omega\mu} V_{\rho+}(\eta,0) - \frac{\alpha_o \eta}{\omega\mu} V_{z+}(\eta,0) = -\left(n + \frac{\tau_o^2}{\omega\mu} Z_f \right) I_{z+}(-m,-\Phi) \tag{4.4.3.5}$$

These equations, again, have the form

$$\begin{vmatrix} \xi 1 & 1 \\ \xi 1 & -1 \end{vmatrix} X_+(\eta) = \begin{vmatrix} d_1 & 0 \\ 0 & d_2 \end{vmatrix} Y_+(-m) \tag{4.4.3.6}$$

where:

$$X_+(\eta) = \begin{vmatrix} V_{z+}(\eta,0) \\ I_{z+}(\eta,0) \\ -\frac{\tau_o^2}{\omega\varepsilon} I_{\rho+}(\eta,0) - \frac{\alpha_o \eta}{\omega\varepsilon} I_{z+}(\eta,0) \\ \frac{\tau_o^2}{\omega\mu} V_{\rho+}(-\eta,0) + \frac{\alpha_o \eta}{\omega\mu} V_{z+}(\eta,0) \end{vmatrix}, \qquad Y_+(-m) = \begin{vmatrix} -\frac{\tau_o}{\omega\varepsilon} I_{\rho+}(-m,\Phi) + \frac{\alpha_o m}{\tau_o \omega\varepsilon} I_{z+}(-m,\Phi) \\ -I_{z+}(-m,\Phi) \\ \frac{\tau_o}{\omega\varepsilon} I_{\rho+}(-m,-\Phi) - \frac{\alpha_o m}{\tau_o \omega\varepsilon} I_{z+}(-m,-\Phi) \\ -I_{z+}(-m,-\Phi) \end{vmatrix} \tag{4.4.3.7}$$

with $d_1 = \begin{vmatrix} \tau_o & 0 \\ 0 & n \end{vmatrix}$ and $d_2 = \begin{vmatrix} \tau_o & 0 \\ 0 & (n + \frac{\tau_o^2}{\omega\mu} Z_f) \end{vmatrix}$ being invertible.

The factorization procedure of the above equation is the same as indicated in section 4.4.2.

In particular, we need to factorize the matrix:

$$\begin{vmatrix} \bar{\xi}_- \bar{D}_{\Sigma-} & 0 \\ 0 & \bar{D}_{\Sigma-} \end{vmatrix} \begin{vmatrix} 1 & \dfrac{1}{\bar{\xi}_-} \bar{f}(\bar{\eta}) \\ \bar{\xi}_- \bar{f}(\bar{\eta}) & 1 \end{vmatrix} \begin{vmatrix} \bar{\xi}_+ \bar{D}_{\Sigma+} & 0 \\ 0 & \bar{D}_{\Sigma+} \end{vmatrix} \tag{4.4.3.8}$$

where:

$$\bar{D}_\Sigma = \begin{vmatrix} \dfrac{1}{\tau_o} & 0 \\ 0 & \dfrac{2k\bar{n} + \tau_o^2 Z_f / Z_o}{2\bar{n}(k\bar{n} + \tau_o^2 Z_f / Z_o)} \end{vmatrix}, \quad \bar{D}_\Delta = \begin{vmatrix} 0 & 0 \\ 0 & \dfrac{2k\bar{n} + \tau_o^2 Z_f / Z_o}{2\bar{n}(k\bar{n} + \tau_o^2 Z_f / Z_o)} \end{vmatrix},$$

$$\bar{f} = \begin{vmatrix} 0 & 0 \\ 0 & \dfrac{\tau_o^2 Z_f / Z_o}{(k\bar{n} + \tau_o^2 Z_f / Z_o)} \end{vmatrix} \tag{4.4.3.9}$$

Again, the process of factorization and decomposition in the $\bar{\eta}$ plane can be simplified by introducing the $\bar{w}$ plane defined by $\bar{\eta} = -\tau_o \cos \bar{w}$ (see Chapter 3 and Appendix 4.A).

4.5. Exact solutions for the impedance wedge problems with the GWHEs written in an alternative form – form #2

As stated in section 4.4, different forms of the GWHEs expand the set of cases that have analytical solutions. Moreover, the form of section 4.4 is the one suitable for the application of the Fredholm factorization method for arbitrary impedance wedges (Daniele and Lombardi 2006; see also Chapter 5).

We recall that the problem under consideration is the diffraction of a plane wave by an arbitrary impedance wedge where the boundary conditions (4.1.10) hold.

Taking into account (3.3.47) and the boundary conditions (4.1.10), after algebraic manipulations, (3.3.57)–(3.3.60) of two angular regions 1 and 2 ($\gamma_a = \gamma_b = \Phi$) present the following form:

$$\begin{vmatrix} \sqrt{\tau_o^2-\eta^2}\,1 & 1 \\ \sqrt{\tau_o^2-\eta^2}\,1 & -1 \end{vmatrix} X_+(\eta) = \begin{vmatrix} 1+\sqrt{\tau_o^2-m^2}\,\dfrac{P_a(m)}{\Delta_a(m)} & 0 \\ 0 & 1+\sqrt{\tau_o^2-m^2}\,\dfrac{P_b(m)}{\Delta_b(m)} \end{vmatrix} Y_+(-m) \quad (4.5.1)$$

where:

$$X_+(\eta) = \begin{vmatrix} V_{z+}(\eta,0) \\ I_{z+}(\eta,0) \\ -\dfrac{\tau_o^2}{\omega\varepsilon} I_{\rho+}(\eta,0) - \dfrac{\alpha_o\eta}{\omega\varepsilon} I_{z+}(\eta,0) \\ \dfrac{\tau_o^2}{\omega\mu} V_{\rho+}(-\eta,0) + \dfrac{\alpha_o\eta}{\omega\mu} V_{z+}(\eta,0) \end{vmatrix}, \qquad Y_+(-m) = \begin{vmatrix} -\dfrac{\tau_o^2}{\omega\varepsilon} I_{\rho+}(-m,\Phi) + \dfrac{\alpha_o m}{\omega\varepsilon} I_{z+}(-m,\Phi) \\ \dfrac{\tau_o^2}{\omega\mu} V_{\rho+}(-m,\Phi) - \dfrac{\alpha m}{\omega\mu} V_{z+}(-m,\Phi) \\ \dfrac{\tau_o^2}{\omega\varepsilon} I_{\rho+}(-m,-\Phi) - \dfrac{\alpha_o m}{\omega\varepsilon} I_{z+}(-m,-\Phi) \\ -\dfrac{\tau_o^2}{\omega\mu} V_{\rho+}(-m,-\Phi) + \dfrac{\alpha_o m}{\omega\mu} V_{z+}(-m,-\Phi) \end{vmatrix}$$

(4.5.2)

The matrices in (4.5.1) are of order 4, 1 is the identity matrix in the space of order 2 and the two polynomials matrices of order 2 $P_{a,b}(m)$ are defined by

$$P_a(m) = \begin{vmatrix} Z_e^a\omega\varepsilon\tau_o^2 & \omega\mu\left[m\alpha_o\left(Z_e^aY_h^a - T_e^aT_h^a\right) + T_e^a\tau_o^2\right] \\ \omega\varepsilon\left[-m\alpha_o\left(Z_e^aY_h^a - T_e^aT_h^a\right) + T_h^a\tau_o^2\right] & Y_h^a\omega\mu\tau_o^2 \end{vmatrix}$$

(4.5.3)

$$P_b(m) = \begin{vmatrix} Z_e^b\omega\varepsilon\tau_o & \omega\mu\left[-m\alpha_o\left(Z_e^bY_h^b - T_e^bT_h^b\right) + T_e^b\tau_o^2\right] \\ \omega\varepsilon\left[m\alpha_o\left(Z_e^bY_h^b - T_e^bT_h^b\right) + T_h^b\tau_o^2\right] & Y_h^b\omega\mu\tau_o^2 \end{vmatrix}$$

(4.5.4)

The scalar polynomials $\Delta_{a,b}(m)$ in (4.5.1) are given by

$$\Delta_a(m) = m^2\alpha_o^2\left(Z_e^aY_h^a - T_e^aT_h^a\right) + m\left(T_e^a - T_h^a\right)\alpha_o\tau_o^2 + \tau_o^4 \quad (4.5.5)$$

$$\Delta_b(m) = m^2\alpha_o^2\left(Z_e^bY_h^b - T_e^bT_h^b\right) - m\left(T_e^b - T_h^b\right)\alpha_o\tau_o^2 + \tau_o^4 \quad (4.5.6)$$

We note that, in the case of symmetric wedges, we have $P_a(m)=P_b(-m)$, $\Delta_a(m)=\Delta_b(-m)$. Furthermore, it is remarkable that, in the case of the normal incidence $\alpha_o=0$, the matrices $P_{a,b}$ and the scalars $\Delta_{a,b}$ are constants.

4.5.1. *Exact factorization with diagonal polynomial matrices* $P_{a,b}(m)$

If $\alpha_o=0$ and the parameter $T_{e,h}^{a,b}$ of (4.5.3)–(4.5.6) related to the anisotropic behavior of the wedge is null, the matrices $P_a(m)/\Delta_a(m)$ and $P_b(m)/\Delta_b(m)$ become diagonal:

$$P_a(m)=\begin{vmatrix} Z_e^a\omega\varepsilon\tau_o^2 & 0 \\ 0 & Y_h^a\omega\mu\tau_o^2 \end{vmatrix}, \qquad \Delta_a(m)=\tau_o^4 \tag{4.5.1.1}$$

$$P_b(m)=\begin{vmatrix} Z_e^b\omega\varepsilon\tau_o^2 & 0 \\ 0 & Y_h^b\omega\mu\tau_o^2 \end{vmatrix}, \qquad \Delta_b(m)=\tau_o^4 \tag{4.5.1.2}$$

Recalling that $n=\sqrt{\tau_o^2-m^2}$, $\xi=\sqrt{\tau_o^2-\eta^2}$, (4.5.1) can be rewritten as

$$\begin{vmatrix} \xi 1 & 1 \\ \xi 1 & -1 \end{vmatrix} X_+(\eta)=\begin{vmatrix} D_a & 0 \\ 0 & D_b \end{vmatrix} Y_+(-m) \tag{4.5.1.3}$$

where the diagonal matrices D_a and D_b are defined by

$$D_a(m)=\begin{vmatrix} 1+n\,Z_e^a\omega\varepsilon\tau_o^{-2} & 0 \\ 0 & 1+n\,Y_h^a\omega\mu\tau_o^{-2} \end{vmatrix},$$
$$D_b(m)=\begin{vmatrix} 1+n\,Z_e^b\omega\varepsilon\tau_o^{-2} & 0 \\ 0 & 1+n\,Y_h^b\omega\mu\tau_o^{-2} \end{vmatrix} \tag{4.5.1.4}$$

Equation (4.5.1.3) becomes

$$\begin{vmatrix} \xi\,D_a^{-1} & D_a^{-1} \\ \xi\,D_b^{-1} & -D_b^{-1} \end{vmatrix} X_+(\eta)=Y_+(-m) \tag{4.5.1.5}$$

We now have

$$\begin{vmatrix} \xi D_a^{-1} & D_a^{-1} \\ \xi D_b^{-1} & -D_b^{-1} \end{vmatrix} = \frac{1}{2} \begin{vmatrix} 1 & 1 \\ 1 & -1 \end{vmatrix} \begin{vmatrix} \xi D_\Sigma & D_\Delta \\ \xi D_\Delta & D_\Sigma \end{vmatrix} \tag{4.5.1.6}$$

where:

$$D_\Sigma = \frac{1}{2}(D_A^{-1} + D_B^{-1}) \quad D_\Delta = \frac{1}{2}(D_A^{-1} - D_B^{-1}) \tag{4.5.1.7}$$

Since D_Σ and D_Δ contain diagonal sub-matrices of order 2, they commute, thus their factorizations

$$D_\Sigma = D_{\Sigma-} D_{\Sigma+} \text{ and } D_\Delta = D_{\Delta-} D_{\Delta+} \tag{4.5.1.8}$$

yield

$$\begin{vmatrix} \xi D_\Sigma & D_\Delta \\ \xi D_\Delta & D_\Sigma \end{vmatrix} = \begin{vmatrix} \xi_- D_{\Sigma-} & 0 \\ 0 & D_{\Sigma-} \end{vmatrix} \begin{vmatrix} 1 & \frac{1}{\xi_-} D_{\Sigma-}^{-1} D_\Delta D_{\Sigma+}^{-1} \\ \xi_- D_{\Sigma-}^{-1} D_\Delta D_{\Sigma+}^{-1} & 1 \end{vmatrix} \begin{vmatrix} \xi_+ D_{\Sigma+} & 0 \\ 0 & D_{\Sigma+} \end{vmatrix} \tag{4.5.1.9}$$

Taking into account that the factorization of $D_\Sigma = D_{\Sigma-} D_{\Sigma+}$ and $D_\Delta = D_{\Delta-} D_{\Delta+}$ involves only the factorization of scalars, the previous expression reduces the problem to the factorization of the matrix of order 4:

$$G_d = \begin{vmatrix} 1 & \frac{1}{\xi_-} f \\ \xi_- f & 1 \end{vmatrix} = \begin{vmatrix} 1 & af \\ bf & 1 \end{vmatrix} \tag{4.5.1.10}$$

where the matrix of order 2 $f = D_{\Sigma-}^{-1} D_\Delta D_{\Sigma+}^{-1}$ is diagonal and $a = 1/\xi_-, b = \xi_-$.

Henceforth, we will define all the quantities in the $\bar{\eta}$ plane where the GWHEs become classical WH equations. For example, the factorization of the scalar

$$\xi = \sqrt{\tau_o^2 - \eta^2} = \sqrt{\tau_o^2 - (-\tau_o \cos[\frac{\Phi}{\pi} \arccos[-\frac{\bar{\eta}}{\tau_o}]]^2} \tag{4.5.1.11}$$

in the $\bar{\eta}$ plane are presented in Daniele and Zich (2014) and in sections 3.8 and 6.3:

$$\bar{\xi}_- = \sqrt{\frac{\tau_o + \bar{\eta}}{2}} \tag{4.5.1.12}$$

This result shows that the ratio

$$\frac{a}{b} = \frac{1}{\xi_-^2} = \frac{2}{\tau_o + \bar{\eta}}$$

is rational so that the matrix G_d can be factorized, as illustrated in section 1.8.

We obtain the following result:

$$G_{d\mp} = \begin{vmatrix} \left[\sqrt{1-f^2}\right]_\mp & 0 \\ 0 & \left[\sqrt{1-f^2}\right]_\mp \end{vmatrix} \begin{Vmatrix} \cosh\left[\frac{1}{2}\xi_- t_\mp\right] & \frac{1}{\xi_-}\sinh\left[\frac{1}{2}\xi_- t_\mp\right] \\ \xi_- \cosh\left[\frac{1}{2}\xi_- t_\mp\right] & \cosh\left[\frac{1}{2}\xi_- t_\mp\right] \end{Vmatrix} \quad (4.5.1.13)$$

where:

$$t_\mp = \left\{\frac{1}{\xi_-}\log\frac{1+f}{1-f}\right\}_\mp \quad (4.5.1.14)$$

As discussed in section 4.4.2, the factorization of the diagonal matrix $\sqrt{1-f^2}$ and the decomposition of the diagonal matrix $t = \frac{1}{\xi_-}\log\frac{1+f}{1-f}$ need to be accomplished in the plane $\bar{\eta} = -\tau_o \cos\bar{w}$. Since f is defined in the *m plane*, we need to apply the substitutions

$$m = \tau_o \cos(w + \Phi) = \tau_o \cos(\frac{\Phi}{\pi}\bar{w} + \Phi) = \tau_o \cos[\frac{\Phi}{\pi}\arccos(-\frac{\bar{\eta}}{\tau_o}) + \Phi] \quad (4.5.1.15)$$

to obtain $\bar{f}(\bar{\eta})$. However, it is convenient to use $\tilde{f}(\bar{w}) = f(\tau_o \cos(\frac{\Phi}{\pi}\bar{w} + \Phi))$ since it is easier to perform decompositions and factorizations in the $\bar{w}$ *plane* according to the equations and properties shown in section 3.3 and Appendix 4.A and in Daniele and Zich (2014).

We also note that the factorized matrices present the algebraic behavior as $\bar{\eta} \to \infty$.

The factorization considered in this section includes, as a particular case, the Malyuzhinets problem, i.e. with $Z_e^a = \frac{1}{Y_h^a}, Z_e^b = \frac{1}{Y_h^b}$.

4.5.2. *Anisotropic symmetric impedance wedges at normal incidence*

Let us consider an anisotropic symmetric impedance wedge at normal incidence. In this case, the matrices

$$h_{a,b} = \begin{bmatrix} Z_e^{a,b} & T_e^{a,b} \\ T_h^{a,b} & Y_h^{a,b} \end{bmatrix} \tag{4.5.2.1}$$

define the boundary conditions on the two faces. Consequently, by imposing $\alpha_o = 0$, the matrices $P_a(m)$ and $P_b(-m)$ and the scalars $\Delta_{a,b}(m)$ are respectively equal and constant:

$$P_a(m) = P_b(-m) = \begin{vmatrix} Z_e^a \omega\varepsilon\tau_o^2 & \omega\mu T_e^a \tau_o^2 \\ \omega\varepsilon T_h^a \tau_o^2 & Y_h^a \omega\mu\tau_o^2 \end{vmatrix}, \qquad \Delta_a(m) = \Delta_b(m) = \Delta = \tau_o^4 \tag{4.5.2.2}$$

Note that $P_a = P_b$ only when $\alpha_o = 0$. Unfortunately, at skew incidence, this property does not hold.

Since $P_a(m) = P_b(-m)$ are constant, their diagonal representation is

$$P_a(m) = P_b(m) = S \begin{vmatrix} d_1 & 0 \\ 0 & d_2 \end{vmatrix} S^{-1} \tag{4.5.2.3}$$

with constant diagonal matrices S and $d = \begin{vmatrix} d_1 & 0 \\ 0 & d_2 \end{vmatrix}$. Taking into account that:

$$\begin{vmatrix} \xi 1 & 1 \\ \xi 1 & -1 \end{vmatrix} = \begin{vmatrix} S & 0 \\ 0 & S \end{vmatrix} \begin{vmatrix} \xi 1 & 1 \\ \xi 1 & -1 \end{vmatrix} \begin{vmatrix} S^{-1} & 0 \\ 0 & S^{-1} \end{vmatrix} \tag{4.5.2.4}$$

the WH equation reduces to

$$\begin{vmatrix} D & 0 \\ 0 & D \end{vmatrix} \begin{vmatrix} \xi 1 & 1 \\ \xi 1 & -1 \end{vmatrix} \tilde{X}_+(\eta) = \tilde{Y}_+(-m) \tag{4.5.2.5}$$

where:

$$\tilde{X}_+(\eta) = \begin{vmatrix} S^{-1} & 0 \\ 0 & S^{-1} \end{vmatrix} X_+(\eta) \tag{4.5.2.6}$$

$$\tilde{Y}_+(-m) = \begin{vmatrix} S^{-1} & 0 \\ 0 & S^{-1} \end{vmatrix} Y_+(-m) \tag{4.5.2.7}$$

$$D^{-1} = \begin{vmatrix} 1+\dfrac{n}{\Delta} & 0 \\ 0 & 1+\dfrac{n}{\Delta} \end{vmatrix} \tag{4.5.2.8}$$

By taking into account that

$$\begin{vmatrix} D & 0 \\ 0 & D \end{vmatrix} \begin{vmatrix} \xi 1 & 1 \\ \xi 1 & -1 \end{vmatrix} = \begin{vmatrix} \xi D & D \\ \xi D & -D \end{vmatrix} = \begin{vmatrix} 1 & 1 \\ 1 & -1 \end{vmatrix} \begin{vmatrix} \xi D & 0 \\ 0 & D \end{vmatrix} \tag{4.5.2.9}$$

the factorization problem is reduced to the factorization of the matrix

$$\begin{vmatrix} \xi D & 0 \\ 0 & D \end{vmatrix} \tag{4.5.2.10}$$

which is diagonal.

We note that the SM solution for this problem was obtained by Lyalinov and Zhu (1999a).

4.5.3. *Non-symmetric wedges at normal incidence with commuting P_a and P_b*

In the presence of wedge anisotropies where $T_{e,h}^{a,b}$ is not null, we can obtain exact solutions, provided that the constant matrices P_a and P_b are commuting. In fact, with this property, they have the same eigenvectors S. The Jordan decomposition yields

$$P_a(m) = SD_aS^{-1}, \; P_b(m) = SD_bS^{-1} \tag{4.5.3.1}$$

where the diagonal matrices D_a and D_b are different. Following the same procedure as in the previous sections, the WH equations again reduce to

$$\begin{vmatrix} D_1 & 0 \\ 0 & D_2 \end{vmatrix} \begin{vmatrix} \xi 1 & 1 \\ \xi 1 & -1 \end{vmatrix} \tilde{X}_+(\eta) = \tilde{Y}_+(-m) \tag{4.5.3.2}$$

where:

$$\tilde{X}_+(\eta) = \begin{vmatrix} S^{-1} & 0 \\ 0 & S^{-1} \end{vmatrix} X_+(\eta) \tag{4.5.3.3}$$

$$\tilde{Y}_+(-m) = \begin{vmatrix} S^{-1} & 0 \\ 0 & S^{-1} \end{vmatrix} Y_+(-m) \tag{4.5.3.4}$$

$$D_1 = (1 + \frac{n}{\Delta_a} D_a)^{-1}, \; D_2 = (1 + \frac{n}{\Delta_a} D_b)^{-1} \tag{4.5.3.5}$$

Using the same algebraic manipulations and the same notations given in the previous sections, the factorization problem is reduced to the factorization of the matrix of order 4:

$$G_d = \begin{vmatrix} 1 & \frac{1}{\xi_-} f \\ \xi_- f & 1 \end{vmatrix} = \begin{vmatrix} 1 & af \\ bf & 1 \end{vmatrix} \tag{4.5.3.6}$$

where the matrix of order 2 $f = D_{\Sigma-}^{-1} D_\Delta D_{\Sigma+}^{-1}$ is diagonal and $a = 1/\xi_-, b = \xi_-$.

Again, the ratio

$$\frac{a}{b} = \frac{1}{\xi_-^2} = \frac{2}{\tau_o + \bar{\eta}} \tag{4.5.3.7}$$

is rational. The matrix G_d can be factorized as illustrated in section 1.8.

Imposing P_a commuting with P_b, when $\alpha_o = 0$, yields the following conditions on the parameters of the wedge:

$$-T_e^b T_h^a + T_e^a T_h^b = 0 \tag{4.5.3.8}$$

$$T_e^b (Z_e^a \varepsilon - Y_h^a \mu) - T_e^a (Z_e^b \varepsilon - Y_h^b \mu) = 0 \tag{4.5.3.9}$$

$$T_h^b (Z_e^a \varepsilon - Y_h^a \mu) - T_h^a (Z_e^b \varepsilon - Y_h^b \mu) = 0 \tag{4.5.3.10}$$

These three equations are satisfied if all T terms are null or also in symmetric wedges. Other cases that yield exact factorizations derive from impedance wedges that satisfy the following four conditions:

$$T_e^a = \pm T_h^a, \quad T_e^b = \pm T_h^b, \quad Z_e^a \varepsilon - Y_h^a \mu = 0, \quad Z_e^b \varepsilon - Y_h^b \mu = 0 \qquad (4.5.3.11)$$

4.5.4. *Non-symmetric wedges at skew incidence*

When studying the normal incidence, we noted that the possibility of obtaining exact factorizations depends on the forms of the matrices P_a and P_b.

In order to obtain exact factorizations at skew incidence $\alpha_o \neq 0$, it is important to study the forms of P_a and P_b. In particular, we study the conditions that make these matrices commute. This study presents several analogies with the work of Bernard (1998) who attempted to obtain a commutative algebra in the matrices arising from the SM method.

To obtain the conditions that make P_a and P_b commute, we consider the commutator K defined by:

$$K(m) = P_a P_b - P_b P_a \qquad (4.5.4.1)$$

Substituting (4.5.3) and (4.5.4) into (4.5.4.1), it yields

$$K(m) = K_o + \alpha_o m K_1 \qquad (4.5.4.2)$$

where K_o and K_1 are constant matrices having the following structure:

$$K_o = \begin{vmatrix} a & b \\ c & -a \end{vmatrix}, \qquad K_1 = \begin{vmatrix} d & e \\ -\dfrac{\mu}{\varepsilon} e & d \end{vmatrix}, \qquad (4.5.4.3)$$

where a, b, c, d and e are constants (which are not reported here). P_a and P_b commute if both K_o and K_1 are null. $K_o = 0$ implies that the three scalars a, b and c are zero. Algebraic manipulations yield (as expected) the same three equations (4.5.3.8)–(4.5.3.10) considered in the case of normal incidence.

The vanishing of K_1 implies that the two scalars d and e are also zero. It yields two additional equations

$$T_e^a T_e^b T_h^a + T_e^a T_e^b T_h^b + T_e^a T_h^a T_h^b + T_e^b T_h^a T_h^b - (T_e^b + T_h^b) Y_h^a Z_e^a - (T_e^a + T_h^a) Y_h^b Z_e^b = 0 \tag{4.5.4.4}$$

$$-(Y_h^a + Y_h^b) Z_e^a Z_e^b \varepsilon + (Z_e^a + Z_e^b) Y_h^a Y_h^b \mu + T_e^b T_h^b (Z_e^a \varepsilon - Y_h^a \mu) + T_e^a T_h^a (Z_e^b \varepsilon - Y_h^b \mu) = 0 \tag{4.5.4.5}$$

Furthermore, the eigenvector matrix $S_{a,b}$ with commuting P_a and P_b is rational (more precisely bilinear) in m if

$$(T_e^a T_h^a - Y_h^a Z_e^a)(Z_e^{a^2} \varepsilon^2 + T_e^{a^2} \varepsilon\mu + 2T_e^a T_h^a \varepsilon\mu + T_h^{a^2} \varepsilon\mu - 2Y_h^a Z_e^a \varepsilon\mu + Y_h^{a^2} \mu^2) = 0 \tag{4.5.4.6}$$

It is remarkable that sufficient conditions to satisfy all the above equations are

$$T_e^a = -T_h^a, \; T_e^b = -T_h^b, \; Z_e^a \varepsilon - Y_h^a \mu = 0, \; Z_e^b \varepsilon - Y_h^b \mu = 0 \tag{4.5.4.7}$$

In particular, with assumptions (4.5.4.7), the eigenvector matrix S of P_a and P_b is constant:

$$S = \begin{vmatrix} \sqrt{-\dfrac{\mu}{\varepsilon}} & -\sqrt{-\dfrac{\mu}{\varepsilon}} \\ 1 & 1 \end{vmatrix} \tag{4.5.4.8}$$

The algebraic manipulations of (1.8.13) to get the factorization of (1.8.19) with the Daniele–Khrapkov form are applied to (4.5.1) when P_a and P_b commute. It yields that the factorization problem is reduced to that of

$$\begin{vmatrix} 1 & \dfrac{1}{\xi_-} f \\ \xi_- f & 1 \end{vmatrix} \tag{4.5.4.9}$$

where:

$$f = P_{\Sigma-}^{-1} P_\Delta P_{\Sigma+}^{-1} = P_\Sigma^{-1} P_\Delta \tag{4.5.4.10}$$

To deduce (4.5.4.9), the following commuting matrices are introduced:

$$P_1 = (1 + \frac{n}{\Delta_a} P_a)^{-1}, \qquad P_2 = (1 + \frac{n}{\Delta_b} P_b)^{-1} \tag{4.5.4.11}$$

$$P_\Sigma = \frac{1}{2}(P_1 + P_2), \qquad P_\Delta = \frac{1}{2}(P_1 - P_2) \tag{4.5.4.12}$$

Furthermore, considering the Jordan form $P_\Sigma = S d_\Sigma S^{-1}$, the factorized matrices of $P_\Sigma = P_{\Sigma-} P_{\Sigma+}$ are the commuting matrices $P_{\Sigma-} = S d_{\Sigma-} S^{-1}$ and $P_{\Sigma+} = S d_{\Sigma+} S^{-1}$, where the diagonal matrix d_Σ is factorized by the diagonal matrices $d_{\Sigma-}$ and $d_{\Sigma+}$, i.e. $d_\Sigma = d_{\Sigma-} d_{\Sigma+}$.

Finally, (4.1.4) and (4.1.5) provide the factorization of (4.5.4.9).

The case of balanced hybrid conditions

$$T_e^a = -T_h^a = 0, \; T_e^b = -T_h^b = 0, \; Z_e^a \varepsilon - Y_h^a \mu = 0, \; Z_e^b \varepsilon - Y_h^b \mu = 0 \tag{4.5.4.13}$$

that assure the commutativity of P_a and P_b, as well as the presence of constant eigenvectors of these matrices, was also completely solved using the SM method in Bernard (1998).

4.5.5. *Two particular wedge problems amenable to exact solutions*

In the study by Lyalinov and Zhu (1999b), the authors used the SM method to solve the diffraction at skew incidence by an impedance wedge with particular face impedances, which, using the current notations, are equivalent to

$$q_{12}^{a,b} = jZ_e^{a,b} - T_e^{a,b} Z_o - T_h^{a,b} Z_o - jY_h^{a,b} Z_o^2 = 0 \tag{4.5.5.1}$$

$$q_{21}^{a,b} = -jZ_e^{a,b} - T_e^{a,b} Z_o - T_h^{a,b} Z_o + jY_h^{a,b} Z_o^2 = 0 \tag{4.5.5.2}$$

In the following, we present how to obtain the exact WH solution of two problems.

In the first problem, the impedances satisfy both constraints (4.5.5.1) and (4.5.5.2), while in the second, we have a few restrictions, i.e. only one of the two constraints is satisfied.

In the first problem, we note that the combination of constraints (4.5.5.1) and (4.5.5.2) is equivalent to (4.5.4.13) given in the previous section. Consequently, the procedure to obtain the WH solution of this problem is already discussed in section 4.5.4. In this section, we provide further considerations on this solution.

The functional equations (3.3.57) and (3.3.58) of angular region 1 can be rewritten in the form:

$$\xi V_{z+}(\eta,0)-\frac{\tau_o^2}{\omega\varepsilon}I_{\rho+}(\eta,0)-\frac{\alpha_o\eta}{\omega\varepsilon}I_{z+}(\eta,0)=-nV_{z+}(-m,\Phi)-\frac{\tau_o^2}{\omega\varepsilon}I_{\rho+}(-m,\Phi)+\frac{\alpha_o m}{\omega\varepsilon}I_{z+}(-m,\Phi) \tag{4.5.5.3}$$

$$\begin{aligned} &jZ_o[\xi\, I_{z+}(\eta,0)+\frac{\tau_o^2}{\omega\mu}V_{\rho+}(-\eta,0)+\frac{\alpha_o\eta}{\omega\mu}V_{z+}(\eta,0)] \\ &=jZ_o[n\;I_{z+}(-m,\Phi)+\frac{\tau_o^2}{\omega\mu}V_{\rho+}(-m,\Phi)-\frac{\alpha_o m}{\omega\mu}V_{z+}(-m,\Phi)] \end{aligned} \tag{4.5.5.4}$$

By adding (4.5.5.3) and (4.5.5.4) and imposing the impedance boundary conditions (4.5.5.5) and (4.5.5.6) on the face a $(\varphi=+\Phi)$

$$V_{\rho+}(-m,\Phi)=-\frac{T_h^a}{T_e^aT_h^a-Y_h^aZ_e^a}V_{z+}(-m,\Phi)+\frac{Z_e^a}{T_e^aT_h^a-Y_h^aZ_e^a}I_{z+}(-m,\Phi) \tag{4.5.5.5}$$

$$I_{\rho+}(-m,\Phi)=-\frac{Y_h^a}{T_e^aT_h^a-Y_h^aZ_e^a}V_{z+}(-m,\Phi)+\frac{T_e^a}{T_e^aT_h^a-Y_h^aZ_e^a}I_{z+}(-m,\Phi) \tag{4.5.5.6}$$

it yields

$$\begin{aligned} \xi X_{1+}+X_{2+}=&\left[-n-\frac{j\,mZ_o\alpha_o}{\mu\omega}+\frac{jT_h^aZ_o\varepsilon\tau_o^2-Y_h^a\mu\tau_o^2}{-(T_e^aT_h^a-Y_h^aZ_e^a)\varepsilon\mu\omega}\right]V_{z+}(-m,\Phi) \\ &+\left[-n-\frac{j\,m\alpha_o}{Z_o\varepsilon\omega}+\frac{jT_e^aY_o\varepsilon\tau_o^2+Z_e^aZ_o\varepsilon\tau_o^2}{(T_e^aT_h^a-Y_h^aZ_e^a)\varepsilon\mu\omega}\right]Z_oI_{z+}(-m,\Phi) \end{aligned} \tag{4.5.5.7}$$

where:

$$X_{1+}=\;V_{z+}(\eta,0)+jZ_oI_{z+}(\eta,0) \tag{4.5.5.8}$$

$$X_{2+}=-\frac{\tau_o^2}{\omega\varepsilon}I_{\rho+}(\eta,0)-\frac{\alpha_o\eta}{\omega\varepsilon}I_{z+}(\eta,0)+\frac{\tau_o^2}{\omega\mu}jZ_oV_{\rho+}(\eta,0)+\frac{\alpha_o\eta}{\omega\mu}jZ_oV_{z+}(\eta,0) \tag{4.5.5.9}$$

Now condition (4.5.5.1) for the face a

$$jZ_e^a-T_e^aZ_o-T_h^aZ_o-jY_h^aZ_o^2=0 \tag{4.5.5.10}$$

makes the two coefficients of $V_{z+}(-m,\Phi)$ and $jZ_oI_{z+}(-m,\Phi)$ equal. Thus, (4.5.5.7) becomes

$$\xi X_{1+} + X_{2+} = \left[-n - \frac{j m Z_o \alpha_o}{\mu\omega} + \frac{j T_h^a Z_o \varepsilon \tau_o^2 - Y_h^a \mu \tau_o^2}{-(T_e^a T_h^a - Y_h^a Z_e^a)\varepsilon\mu\omega}\right] Y_{1-} \tag{4.5.5.11}$$

where:

$$Y_{1-} = V_{z+}(-m,\Phi) + jZ_o I_{z+}(-m,\Phi) \tag{4.5.5.12}$$

The application of condition (4.5.5.1) for the face b

$$jZ_e^b - T_e^b Z_o - T_h^b Z_o - jY_h^b Z_o^2 = 0 \tag{4.5.5.13}$$

to (3.3.59) and (3.3.60) of angular region 2 yields

$$\xi X_{1+} - X_{2+} = \left[-n + \frac{j m Z_o \alpha_o}{\mu\omega} + \frac{j T_h^b Z_o \varepsilon \tau_o^2 - Y_h^b \mu \tau_o^2}{-(T_e^b T_h^b - Y_h^b Z_e^b)\varepsilon\mu\omega}\right] Y_{2-} \tag{4.5.5.14}$$

with

$$Y_{2-} = V_{z+}(-m,-\Phi) + jZ_o I_{z+}(-m,-\Phi) \tag{4.5.5.15}$$

Equations (4.5.5.11) and (4.5.5.14) constitute two coupled WH equations that involve a matrix kernel of order 2 having the form

$$\begin{vmatrix} \xi 1 & 1 \\ \xi 1 & -1 \end{vmatrix} X_+(\eta) = \begin{vmatrix} d_a & 0 \\ 0 & d_b \end{vmatrix} Y_+(-m) \tag{4.5.5.16}$$

where the scalars $d_{a,b}$ are given by

$$d_a = -n - \frac{j m Z_o \alpha_o}{\mu\omega} + \frac{j T_h^a Z_o \varepsilon \tau_o^2 - Y_h^a \mu \tau_o^2}{-(T_e^a T_h^a - Y_h^a Z_e^a)\varepsilon\mu\omega} \tag{4.5.5.17}$$

$$d_b = -n + \frac{j m Z_o \alpha_o}{\mu\omega} + \frac{j T_h^b Z_o \varepsilon \tau_o^2 - Y_h^b \mu \tau_o^2}{-(T_e^b T_h^b - Y_h^b Z_e^b)\varepsilon\mu\omega} \tag{4.5.5.18}$$

Following the usual procedure given in section 1.8, it yields to a matrix of the Daniele–Khrapkov type that can be factorized in closed form. Thus, we can compute the solution in terms of the unknowns Y_{1-} and Y_{2-}, as well as X_{1+} and X_{2+}.

The knowledge of X_{1+} and X_{2+} provides only two relations (4.5.5.8) and (4.5.5.9) for the computation of the four plus spectra $V_{z+}(\eta,0)$, $I_{z+}(\eta,0)$, $V_{\rho+}(\eta,0)$, $I_{\rho+}(\eta,0)$. In order to solve the problem, we need other two equations, and therefore we use equation (4.5.5.2) (which until now has not been used).

By subtracting first (3.3.57) and (3.3.58) and successively (3.3.59) and (3.3.60), taking into account (4.5.5.2) and applying a procedure similar to that for (4.5.5.11) and (4.5.5.14), we again obtain two uncoupled WH equations that can be factorized in closed form and provide the two plus functions X_{3+} and X_{4+}

$$X_{3+} = V_{z+}(\eta,0) - jZ_o I_{z+}(\eta,0), \tag{4.5.5.19}$$

$$X_{4+} = -\frac{\tau_o^2}{\omega\varepsilon} I_{\rho+}(\eta,0) - \frac{\alpha_o\eta}{\omega\varepsilon} I_{z+}(\eta,0) - \frac{\tau_o^2}{\omega\mu} jZ_o V_{\rho+}(\eta,0) - \frac{\alpha_o\eta}{\omega\mu} jZ_o V_{z+}(\eta,0) \tag{4.5.5.20}$$

The knowledge of X_{1+}, X_{2+}, X_{3+} and X_{4+} provides enough relations to compute the four plus spectra $V_{z+}(\eta,0)$, $I_{z+}(\eta,0)$, $V_{\rho+}(\eta,0)$, $I_{\rho+}(\eta,0)$. It completes the solution of the problem.

In a successive paper by Lyalinov and Zhu (2003b), the same authors solved the wedge problem when the impedance conditions on the faces only satisfy condition (4.5.5.1) $q_{12}^{a,b} = 0$ and not (4.5.5.2) $q_{21}^{a,b} \neq 0$.

This means that while (4.5.5.11) and (4.5.5.14) still hold, i.e. X_{1+}, Y_{1-} and Y_{2-} X_{2+} are known, we cannot use the second part of the previously illustrated procedure, which allowed the computation of X_{3+}, X_{4+} and the corresponding minus functions Y_{3-}, Y_{4-}.

However, we can rewrite (3.3.57) and 3.3.58) in the form

$$\xi V_{z+}(\eta,0) - \frac{\tau_o^2}{\omega\varepsilon} I_{\rho+}(\eta,0) - \frac{\alpha_o\eta}{\omega\varepsilon} I_{z+}(\eta,0)$$
$$= \left[jZ_o n + \frac{m\alpha_o}{\varepsilon\omega} - \frac{\left(T_e^a + jY_h^a Z_o\right)\tau_o^2}{\left(T_e^a T_h^a - Y_h^a Z_e^a\right)\varepsilon\omega} \right] I_{z+}(-m,\Phi) + \left[n - \frac{jY_h^a \tau_o^2}{\left(T_e^a T_h^a - Y_h^a Z_e^a\right)\varepsilon\omega} \right] Y_{1-} \tag{4.5.5.21}$$

$$\xi V_{z+}(\eta,0)+\frac{\tau_0^2}{\omega\varepsilon}I_{\rho+}(\eta,0)+\frac{\alpha_0\eta}{\omega\varepsilon}I_{z+}(\eta,0)$$

$$=\left[jZ_o n-\frac{m\alpha_o}{\varepsilon\omega}-\frac{\left(T_e^b+jY_h^b Z_o\right)\tau_o^2}{\left(T_e^b T_h^b-Y_h^b Z_e^b\right)\varepsilon\omega}\right]I_{z+}(-m,-\Phi)+\left[-n+\frac{jY_h^b\tau_o^2}{\left(T_e^b T_h^b-Y_h^b Z_e^b\right)\varepsilon\omega}\right]Y_{2-} \quad (4.5.5.22)$$

Again, we deal with two coupled WH equations that involve a matrix kernel of the Daniele–Khrapkov type. However, there is now the presence of two inhomogeneous terms

$$S_1=[n-\frac{jY_h^a\tau_o^2}{(T_e^a T_h^a-Y_h^a Z_e^a)\varepsilon\omega}]Y_{1-} \quad (4.5.5.23)$$

$$S_2=[-n+\frac{jY_h^b\tau_o^2}{(T_e^b T_h^b-Y_h^b Z_e^b)\varepsilon\omega}]Y_{2-} \quad (4.5.5.24)$$

It is important to note that quantities (4.5.5.23) and (4.5.5.24) are known from the solutions of (4.5.5.11) and (4.5.5.14). The solution of the inhomogeneous WH equation in the presence of sources S_1 and S_2, which are not usual plane wave contributions, requires non-simple Cauchy decomposition formulas.

We can accomplish this decomposition in the $\bar{w}$ *plane*, as illustrated in section 3.3 of Daniele and Zich (2014).

We note that, in this case, the decomposition formulas have the same role of the S-integrals in the SM technique.

The solution of homogeneous equations (4.5.5.11) and (4.5.5.14) and successively the solution of inhomogeneous equations (4.5.5.21) and (4.5.5.22) complete the WH solution of the problem studied using the SM technique in Lyalinov and Zhu (2003b).

4.6. A general form of the GWHEs to study the arbitrary face impedance wedges – form #3

In this section, we deduce the general form (4.1.1) of the matrices to study the diffraction by impedance wedges using factorization.

By multiplying both the members of (4.5.1) by

$$\begin{vmatrix} 1-\sqrt{\tau_o^2-m^2}\,\dfrac{P_a(m)}{\Delta_a(m)} & 0 \\ 0 & 1-\sqrt{\tau_o^2-m^2}\,\dfrac{P_b(m)}{\Delta_b(m)} \end{vmatrix} \tag{4.6.1}$$

it yields the GWHEs

$$G_\Phi(\eta)X_+(\eta)=Y_+(-m) \tag{4.6.2}$$

where:

$$G_\Phi(\eta)=R_1^{-1}(m)\cdot\begin{vmatrix} 1-n\dfrac{P_a(m)}{\Delta_a(m)} & 0 \\ 0 & 1-n\dfrac{P_b(m)}{\Delta_b(m)} \end{vmatrix}\cdot\begin{vmatrix} \xi 1 & 1 \\ \xi 1 & -1 \end{vmatrix} \tag{4.6.3}$$

with $n=\sqrt{\tau_o^2-m^2}$, $\xi=\sqrt{\tau_o^2-\eta^2}$, and the rational matrix $R_1\,(m)$ is given by

$$R_1\,(m)=\begin{vmatrix} 1-n^2\dfrac{P_a^2(m)}{\Delta_a^2(m)} & 0 \\ 0 & 1-n^2\dfrac{P_b^2(m)}{\Delta_b^2(m)} \end{vmatrix} \tag{4.6.4}$$

Now, considering in (4.6.3)

$$\begin{vmatrix} 1-n\dfrac{P_a(m)}{\Delta_a(m)} & 0 \\ 0 & 1-n\dfrac{P_b(m)}{\Delta_b(m)} \end{vmatrix}\cdot\begin{vmatrix} \xi 1 & 1 \\ \xi 1 & -1 \end{vmatrix}=\begin{vmatrix} \xi\left(1-n\dfrac{P_a(m)}{\Delta_a(m)}\right) & 1-n\dfrac{P_a(m)}{\Delta_a(m)} \\ \xi\left(1-n\dfrac{P_b(m)}{\Delta_b(m)}\right) & -\left(1-n\dfrac{P_b(m)}{\Delta_b(m)}\right) \end{vmatrix} \tag{4.6.5}$$

it allows us to rewrite (4.6.2)

$$\begin{vmatrix} \xi\left(1-n\dfrac{P_a(m)}{\Delta_a(m)}\right) & 1-n\dfrac{P_a(m)}{\Delta_a(m)} \\ \xi\left(1-n\dfrac{P_b(m)}{\Delta_b(m)}\right) & -\left(1-n\dfrac{P_b(m)}{\Delta_b(m)}\right) \end{vmatrix} X_+(\eta)=R_1(m)Y_+(-m) \tag{4.6.6}$$

which, after algebraic manipulations, becomes

$$\begin{vmatrix} \xi n & r(m)s(m) \\ \xi r(m)s(m) & n \end{vmatrix} X_+(\eta) = r(m)\begin{vmatrix} 1 & -1 \\ 1 & 1 \end{vmatrix} R_1(m)Y_+(-m) \tag{4.6.7}$$

with

$$r(m) = \left(\frac{P_b(m)}{\Delta_b(m)} - \frac{P_a(m)}{\Delta_a(m)}\right)^{-1}, \qquad s(m) = 2 - n\left(\frac{P_b(m)}{\Delta_b(m)} + \frac{P_a(m)}{\Delta_a(m)}\right) \tag{4.6.8}$$

The factorization of the scalars $\bar{\xi} = \bar{\xi}_-\bar{\xi}_+$ and $\bar{n} = \bar{n}_-\bar{n}_+$ in the $\bar{\eta}$ *plane* yields

$$\begin{vmatrix} \bar{\xi}_-\bar{n}_- & 0 \\ 0 & n_- \end{vmatrix} \begin{vmatrix} 1 & \frac{1}{\bar{\xi}_-}f(m) \\ \bar{\xi}_- f(m) & 1 \end{vmatrix} \begin{vmatrix} \bar{\xi}_+\bar{n}_+ & 0 \\ 0 & n_+ \end{vmatrix} X_+(\eta) = R(m)Y_+(-m) \tag{4.6.9}$$

where:

$$f(m) = \frac{1}{n}r(m)s(m) \tag{4.6.10}$$

and the rational matrix R(m) is given by

$$R(m) = r(m)\begin{vmatrix} 1 & -1 \\ 1 & 1 \end{vmatrix} R_1(m) \tag{4.6.11}$$

Although the general exact factorization problem remains very difficult to complete, we note that particular importance is assumed by the classical factorization of the matrix of order 4:

$$\begin{vmatrix} 1 & \frac{1}{\bar{\xi}_-}f(m) \\ \bar{\xi}_- f(m) & 1 \end{vmatrix} = \begin{vmatrix} 1 & \frac{1}{\bar{\xi}_-}\bar{f}(\bar{\eta}) \\ \bar{\xi}_- \bar{f}(\bar{\eta}) & 1 \end{vmatrix} \tag{4.6.12}$$

in particular when studied in the $\bar{\eta}$ *plane*.

The form of the matrix (4.6.12) coincides with that of (4.1.1).

We recall that (4.6.12) is amenable to the exact factorization when $\overline{f}(\overline{\eta})$ is diagonal or, more generally, when its eigenvectors are constants, for example when $\overline{f}(\overline{\eta})$ commutes with a constant matrix.

Appendix 4.A. Some important formulas of decomposition for wedge problems

In wedge problems, the factorization and decomposition of scalars in the $\overline{\eta}$ *plane* can be more easily obtained using the decomposition formula (4.A.3) and (4.A.4) (for the decomposed plus) and (4.A.5) and (4.A.6) (for the decomposed minus), as described in section 3.3 of Daniele and Zich (2014) starting from the classical decomposition formulas described in section 1.6.

These formulas, which are very useful, work in the $\overline{w}$ *plane* ($\overline{\eta} = -\tau_o \cos \overline{w}$; see section 3.4). For example, using them, the factorization of

$$\xi = \sqrt{\tau_o^2 - \eta^2} = \sqrt{\tau_o^2 - (-\tau_o \cos[\frac{\Phi}{\pi} \arccos[-\frac{\overline{\eta}}{\tau_o}]]^2} \tag{4.A.1}$$

yields the factorized scalars

$$\xi_- = \sqrt{\frac{\tau_o + \overline{\eta}}{2}} = \sqrt{\tau_o} \sin \frac{\pi w}{2\Phi}, \qquad \xi_+ = -\sqrt{\tau_o} \frac{\sin w}{\sin \dfrac{\pi w}{2\Phi}} \tag{4.A.2}$$

which are the same expressions obtained by Daniele (2003b), as well as described in sections 3.8 and 6.3 of this book. In fact, (4.A.2) can be derived by exploiting properties of plus and minus functions in $\overline{w}$ *plane* as reported in section 3.4 after (3.4.9).

The decomposition formulas are obtained in both the $\overline{\eta}$ *plane* and the $\overline{w} = \dfrac{\pi}{\Phi} w$ *plane*, where we define $\overline{X}(\overline{\eta}) = \tilde{X}(\overline{w})$. The decomposition formulas are

$$\hat{X}_+(w) = -\frac{1}{2\pi j} \int_0^\infty \frac{[\tilde{X}(-\pi + ju) - \tilde{X}(-\pi - ju)] \sinh u}{\cosh u + \cos\left(\dfrac{\pi}{\Phi} w\right)} du, \tag{4.A.3}$$

or equivalently

$$\bar{X}_+(\bar{\eta}) = -\frac{1}{2\pi j}\int_0^\infty \frac{[\tilde{X}(-\pi+ju)-\tilde{X}(-\pi-ju)]\sinh u}{\cosh u - \dfrac{\bar{\eta}}{\tau_o}}du, \tag{4.A.4}$$

and

$$\hat{X}_-(w) = \frac{1}{2\pi j}\int_0^\infty \frac{[\tilde{X}(ju)-\tilde{X}(ju)]\sinh u}{\cosh u - \cos\left(\dfrac{\pi}{\Phi}w\right)}du, \tag{4.A.5}$$

or equivalently

$$\bar{X}_-(\bar{\eta}) = \frac{1}{2\pi j}\int_0^\infty \frac{[\tilde{X}(ju)-\tilde{X}(ju)]\sinh u}{\cosh u + \dfrac{\bar{\eta}}{\tau_o}}du, \tag{4.A.6}$$

where $\bar{\eta} = -\tau_o \cos\bar{w}$ and $\bar{w} = -\pi/2 + ju$ (see section 3.4).

Note that the decomposition could involve singularities of the integrands that are located in particular regions of the proper sheet of $\bar{\eta}$, as defined in section 3.4 of this book and in section 3.3 of Daniele and Zich (2014). In this case, the formulas must take into account of these singularities.

To show the importance of these formulas, in the Malyuzhinets problem (section 4.4.1), we obtain the decomposition of

$$t\ (\bar{\eta})k = \frac{1}{2\sqrt{k\xi_-}}\log\frac{\bar{n}+n_b}{\bar{n}+n_a} \tag{4.A.7}$$

and the factorization of

$$\bar{g}(\bar{\eta}) = \frac{1}{\sqrt{(\bar{n}+n_a)(\bar{n}+n_b)}} \tag{4.A.8}$$

We must locate in the w *plane* or the $\bar{w}$ *plane* the zeros of $n+n_{a,b}$ to ascertain whether (4.A.3)–(4.A.6) require additive correction contributions due to these zeros. We recall that $n_{a,b} = \tau_o \sin\vartheta_{a,b}$ with $\mathrm{Re}[\sin\vartheta_{a,b}] \geq 0$ or $0 \leq \mathrm{Re}\,\vartheta_{a,b} \leq \pi$.

The possible additive correction contributions derive from the singularities:

$$\frac{n}{\tau_o} + \sin\vartheta = 0 \tag{4.A.9}$$

which are present in the functions $\bar{t}(\bar{\eta})$ and $\bar{g}(\bar{\eta})$.

In the $\bar{w}$ *plane*, the zeros of $n + n_{a,b}$ are located in

$$\sin[\frac{\Phi}{\pi}\bar{w} + \Phi] + \sin\vartheta_{a,b} = 0 \tag{4.A.10}$$

thus we have zeros for

$$\bar{w} = -\frac{\pi}{\Phi}\vartheta_{a,b} - \pi + \frac{2\pi^2}{\Phi}l \quad , \; l = 0, \pm 1, \pm 2, ... \tag{4.A.11}$$

Taking into account that $\Phi < \pi$ and $0 \le \mathrm{Re}\,\vartheta_{a,b} \le \pi$, none of the values $l = 0, \pm 1, \pm 2, ...$ produce zeros located in the proper sheet of the $\bar{w}$ *plane*:

$$-\pi < \mathrm{Re}[\bar{w}] < 0 \tag{4.A.12}$$

This means that, in this case, no correction terms are needed in (4.A.3)–(4.A.6).

We note that the zeros of $n + n_{a,b}$ produce structural poles that have a physical meaning. They represent possible surface waves on the faces of the wedge. Thus, their occurrence is possible when we evaluate the total field for an arbitrary direction φ, as described in section 3.10 (see also section 5.7 for the theory and section 5.10 for an application).

The factorization of (4.4.1.17) can be obtained by considering the logarithmic decomposition of $\bar{X}(\bar{\eta})$

$$\bar{X}(\bar{\eta}) = \log\frac{n + n_o}{n} = \log\frac{\sin[\frac{\Phi}{\pi}\bar{w} + \Phi] + \sin\vartheta}{\sin[\frac{\Phi}{\pi}\bar{w} + \Phi]} \tag{4.A.13}$$

i.e.

$$\tilde{X}(-\pi + ju) = \log\frac{\sinh[\frac{\Phi}{\pi}u] - j\sin\vartheta}{\sinh[\frac{\Phi}{\pi}u]}, \qquad \tilde{X}(-\pi - ju) = \log\frac{\sinh[\frac{\Phi}{\pi}u] + j\sin\vartheta}{\sinh[\frac{\Phi}{\pi}u]} \tag{4.A.14}$$

Using (4.A.14), we obtain

$$\tilde{X}(-\pi + ju) - \tilde{X}(-\pi - ju) = \log\frac{\sinh[\frac{\Phi}{\pi}u] - j\sin\vartheta}{\sinh[\frac{\Phi}{\pi}u] + j\sin\vartheta} = \frac{2}{j}\arctan\frac{\sin\vartheta}{\sinh[\frac{\Phi}{\pi}u]}. \tag{4.A.15}$$

which yields, using (4.A.4), the factorized plus

$$\left[\frac{n+n_o}{n}\right]_+\Bigg|_{\eta=-\tau_o\sin w} = \tilde{d}_\Phi(w) = \exp\left[\frac{1}{\pi}\int_0^\infty \arctan[\frac{\sin\vartheta}{\sinh[\frac{\Phi}{\pi}u]}]\frac{\sinh u}{\cosh u + \cos\left(\frac{\pi}{\Phi}w\right)}du\right] \tag{4.A.16}$$

According to the properties of the plus functions, this function is even in w (see section 1.14). Furthermore, it is real for real values of ϑ.

The same procedure applies for the decomposition of (4.4.1.13). We obtain

$$\bar{t}_+(\bar{\eta}) = \left\{\frac{1}{\bar{\xi}_-(\bar{\eta})}\log[\frac{n+n_b}{n+n_a}]\right\}_+$$
$$= -\frac{2}{\pi}\int_0^\infty \left[\arctan[\frac{\sin\vartheta_b}{\sinh[\frac{\Phi}{\pi}u]}] - \arctan[\frac{\sin\vartheta_a}{\sinh[\frac{\Phi}{\pi}u]}]\right]\frac{\sinh\frac{u}{2}}{\cosh u - \bar{\eta}/k}du \tag{4.A.17}$$

An alternative factorization of $\frac{n+n_o}{n}$ can be obtained in the *w plane* using difference equations derived from the WH formulation.

In fact, the solution of the Wiener-Hopf equation can be often reduced to the solution of the difference equation.

In the *w plane*, we write

$$\frac{n+n_o}{n}=\frac{\sin[w+\Phi]+\sin\vartheta}{\sin[w+\Phi]}=\hat{Y}_+(-w-\Phi)\hat{X}_+(w)=\hat{Y}_+(w+\Phi)\hat{X}_+(w) \tag{4.A.18}$$

or equivalently

$$\frac{\sin[w]+\sin\vartheta}{\sin[w]}=\hat{Y}_+(w)\hat{X}_+(w-\Phi) \tag{4.A.19}$$

Replacing w with $-w$ and taking into account the even symmetry of the plus functions in the *w plane*, we have:

$$\frac{\sin[-w]+\sin\vartheta}{\sin[-w]}=\hat{Y}_+(-w)\hat{X}_+(-w-\Phi)=\hat{Y}_+(w)\hat{X}_+(w+\Phi) \tag{4.A.20}$$

By eliminating $\hat{Y}_+(w)$, we obtain the difference equation of first order

$$\hat{X}_+(w+\Phi)=\frac{\sin[w]-\sin\vartheta}{\sin[w]+\sin\vartheta}\hat{X}_+(w-\Phi) \tag{4.A.21}$$

The solution of the previous equations produces multiple solutions. For example, if $X(w)$ is a solution, $X(w)f(w)$ is also a solution, where $f(w)$ is an arbitrary periodic function with period 2Φ. The correct solution among those that satisfy the difference equation (4.A.21) must present additional properties related to the factorized function $X_+(\eta)=\overline{X}_+(\overline{\eta})=\hat{X}_+(w)=\tilde{X}_+(\overline{w})$.

In particular, $\hat{X}_+(w)$ must be even: $\hat{X}_+(-w)=\hat{X}_+(w)$. Furthermore, since there are no singularities in $-\pi<\mathrm{Re}[\overline{w}]<0$, $\hat{X}_+(w)$ is regular in the strip $-\Phi<\mathrm{Re}[w]<0$ and bounded in the strip for $w\to\pm j\infty$.

The solutions of the difference equation of first order can be obtained by rewriting (4.A.21) as

$$\log[\hat{X}_+(w+\Phi)]-\log[\hat{X}_+(w-\Phi)]=\log\frac{\sin[w]-\sin\vartheta}{\sin[w]+\sin\vartheta} \tag{4.A.22}$$

or by introducing the logarithmic derivative

$$Z(w) = \frac{d}{dw}\log[\hat{X}_+(w)] = -Z(-w) \tag{4.A.23}$$

yielding

$$Z(w+\Phi)] - Z(w-\Phi)] = \frac{d}{dw}\log\frac{\sin[w]-\sin\vartheta}{\sin[w]+\sin\vartheta} = \frac{1}{\sin(w-\vartheta)} - \frac{1}{\sin(w+\vartheta)} = \Delta(w) \tag{4.A.24}$$

Thus, we reduce the problem to the solution of a difference equation with constant coefficients. This equation can be solved using the Malyuzhinets technique (see Appendix 3.A) by introducing the MF transform

$$z(t) = -j\int_{-j\infty}^{j\infty} Z(w)\exp[+jtw]dw \tag{4.A.25}$$

Taking into account that $Z(w)$ is regular in $-\Phi < \mathrm{Re}[w] < +\Phi$, we have:

$$-j\int_{-j\infty}^{j\infty} Z(w+\Phi)\exp[+jtw]dw = z(t)e^{-jt\Phi} \tag{4.A.26}$$

$$-j\int_{-j\infty}^{j\infty} Z(w-\Phi)\exp[+jtw]dw = z(t)e^{+jt\Phi} \tag{4.A.27}$$

It yields

$$e^{-jt\Phi}z(t) - e^{+jt\Phi}z(t) = -j\int_{-j\infty}^{j\infty} \Delta(w)\exp[+jtw]dw = -2\pi\frac{\cos[t(\frac{\pi}{2}-\vartheta)]}{\cos[t\frac{\pi}{2}]} \tag{4.A.28}$$

where the integral is evaluated using the residue theorem.

It follows that:

$$z(t) = \pi\frac{\cos[t(\frac{\pi}{2}-\vartheta)]}{j\sin(\Phi t)\cos[t\frac{\pi}{2}]} \tag{4.A.29}$$

The inverse transform

$$Z(w) = -\frac{j}{2\pi}\int_{-j\infty}^{j\infty} z(t)\exp[-jtw]dt = -\int_{-j\infty}^{j\sigma} \frac{\cos\left[t\left(\frac{\pi}{2}-\vartheta\right)\right]}{2\sin(\Phi t)\cos\left[t\frac{\pi}{2}\right]}\exp[-jtw]dt$$

$$= \int_{-j\infty}^{j\infty} \frac{\cos\left[t\left(\frac{\pi}{2}-\vartheta\right)\right]\cos(\Phi t)}{\sin(2\Phi t)\cos\left[t\frac{\pi}{2}\right]}\exp[+jtw]dt$$

(4.A.30)

yields

$$Z(w)$$

$$= \frac{1}{2}\int_{-j\infty}^{j\omega}\left[\frac{\cos\left[t\left(\frac{\pi}{2}-\vartheta\right)\right]\exp(j\Phi t)}{\sin(2\Phi t)\cos\left[t\frac{\pi}{2}\right]} + \frac{\cos\left[t\left(\frac{\pi}{2}-\vartheta\right)\right]\exp(-j\Phi t)}{\sin(2\Phi t)\cos\left[t\frac{\pi}{2}\right]}\right]\exp[+jtw]dt =$$

$$Z(w)$$

$$= \eta_\Phi\left(w+\Phi+\frac{\pi}{2}-\vartheta\right) + \eta_\Phi\left(w+\Phi-\frac{\pi}{2}+\vartheta\right) + \eta_\Phi\left(w-\Phi+\frac{\pi}{2}-\vartheta\right) + \eta_\Phi\left(w-\Phi-\frac{\pi}{2}+\vartheta\right)$$

(4.A.31)

where the following odd function is introduced:

$$\eta_\Phi(w) = \frac{1}{4}\int_{-j\infty}^{j\infty} \frac{\exp[+jtw]}{\sin(2\Phi t)\cos[t\frac{\pi}{2}]}dt = -\frac{1}{2}\int_0^\infty \frac{\sinh[sw]}{\sinh(2\Phi s)\cosh[s\frac{\pi}{2}]}ds \qquad (4.A.32)$$

It follows that:

$$\hat{X}_+(w) = \exp\left[\int_0^w Z(w)dw\right] = \Psi_\Phi\left(w+\Phi+\frac{\pi}{2}-\vartheta\right)$$

$$+\Psi_\Phi\left(w+\Phi-\frac{\pi}{2}+\vartheta\right) + \Psi_\Phi\left(w-\Phi+\frac{\pi}{2}-\vartheta\right) + \Psi_\Phi\left(w-\Phi-\frac{\pi}{2}+\vartheta\right)$$

(4.A.33)

where the even function

$$\Psi_{\Phi}(w) = \exp\left(\int_0^w \eta_{\Phi}[u]du\right)$$

$$= \exp\left[-\frac{1}{2}\int_0^w\int_0^\infty \frac{\sinh[su]}{\sinh(2\Phi s)\cosh\left[s\frac{\pi}{2}\right]}dsdu\right] = \exp\left[\frac{1}{2}\int_0^\infty \frac{1-\cosh[sw]}{s\cosh\left[s\frac{\pi}{2}\right]\sinh(2\Phi s)}ds\right]$$

(4.A.34)

is the Malyuzhinets function (Senior and Volakis 1995b).

Since the factorized functions are defined within a constant, it follows the relation (4.A.35) that is the link between (4.A.16) and the Malyuzhinets function (4.A.34) and used as given in (4.4.1.26):

$$\hat{X}_+(w) = \frac{\hat{X}_+(0)}{d_{\Phi}(0)} d_{\Phi}(w) \tag{4.A.35}$$

A similar deduction can be used to show the link (4.4.1.27) between (4.A.17) and the Malyuzhinets function (4.A.34).

The use of $d_{\Phi}(w)$ is also discussed in section 5.9 to compare the WH solution of the Malyuzhinets problem reported in section 4.4.1 with that obtained by applying the Fredholm factorization method.

5

Fredholm Factorization Solutions of GWHEs for the Electromagnetic Impedance Wedges Surrounded by an Isotropic Medium

5.1. Introduction

While, in Chapter 3, we presented the general spectral methodologies to study the scattering problem constituted of angular regions illuminated by plane waves in general wave motion modeled by a linear operator, in this chapter, we present the application of the Fredholm factorization method to GWHEs for the solution of the general electromagnetic scattering problem, constituted of an impenetrable wedge immersed in a homogeneous medium.

In particular, this methodology is useful in all the cases when a classical factorization method is not available (a vast set of solutions is reported in Chapter 4).

The most general problem is the one where the impenetrable wedge is modeled by anisotropic non-symmetric impedances faces and illuminated at skew incidence (Daniele 2003b; Daniele and Lombardi 2005, 2006).

In this context, the Fredholm factorization method is always applicable, and in particular, it can solve problems where no other methods allow us to obtain an exact solution (Wiener-Hopf method, Kontorovich–Lebedev method, Sommerfeld–Malyuzhinets, etc.).

This book consists of two volumes; any references to Chapters 1, 2 or 3 can be found in the first volume: *Scattering and Diffraction by Wedges 1*.

The Fredholm factorization method essentially eliminates one of the two unknown kinds (plus or minus) yielding a Fredholm integral equation of the second kind (FIE) with one type of unknowns. In this book, we prefer to keep the plus unknown; however, generalization to keep the minus unknown is straightforward. The technique is developed both in the modified Laplace domain $\bar{\eta}$ and the modified angular complex plane $\bar{w}$, demonstrating the benefits of using $\bar{w}$ to enhance convergence and the advantages of the Laplace domain η to investigate more complex problems with multiple interfaces and different types of regions (angular, rectangular, etc.).

In this chapter, we validate the method for an impenetrable wedge by comparison with other independent solutions, and we propose some interesting practical engineering/physical results in electromagnetic scattering by impenetrable wedge.

In particular, the scope of this chapter is to present the technique with all the details to enable independent implementation by the reader using technical computing software, in particular, in analytic and numerical manipulations of the equations. We hope that the material is consistent, making reference to the theory of previous chapters as well as of sufficient interest for the autonomous development of new technical results in practical applications. For readability, we recall the notation introduced in Chapter 3 for quantities defined in different complex planes, i.e. $f_+(\eta) = f_+(-\tau_o \cos w) = \hat{f}_+(w) = \bar{f}_+(\bar{\eta}) = \bar{f}_+(-\tau_o \cos \bar{w}) = \tilde{f}_+(\bar{w})$.

5.2. Generalized Wiener-Hopf equations for the impenetrable wedge scattering problem of an electromagnetic plane wave at skew incidence

Figure 5.2.1 presents the scattering problem where Cartesian coordinates, together with cylindrical coordinates, are used to describe the problem. We assume the wedge with cylindrical symmetry with respect to the z axis. For simplicity and without loss of generality, we orient the coordinate systems such that the x axis bisects the aperture angle ($2\pi - 2\Phi$) of the wedge.

Following the description within Chapter 3, in this section, we present the generalized Wiener-Hopf equations (GWHEs) for the electromagnetic scattering of an impenetrable wedge with general anisotropic non-symmetric surface impedances (5.2.1) in a homogeneous medium with propagation constant k and impedance Z_o. In order to meet the mathematical requirements of the WH technique, small vanishing losses are assumed in the medium (say $k = k' - jk'', k'' << k'$):

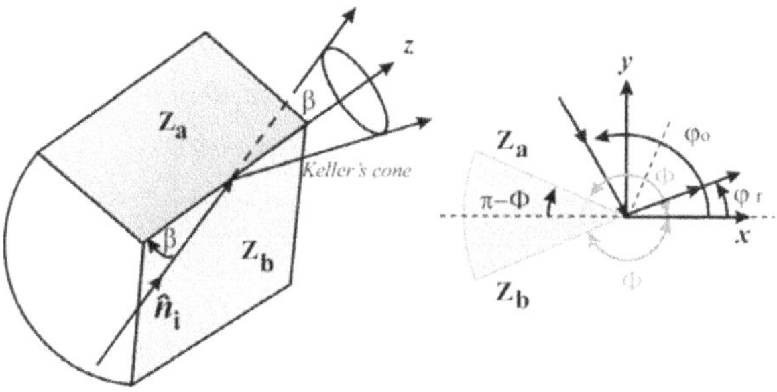

Figure 5.2.1. *Scattering of an impenetrable wedge by a plane wave at skew incidence (face a at $\varphi = \Phi$ and face b at $\varphi = -\Phi$)*

$$\begin{bmatrix} E_z(\rho,\Phi) \\ E_\rho(\rho,\Phi) \end{bmatrix} = Z_a \begin{bmatrix} H_\rho(\rho,\Phi) \\ -H_z(\rho,\Phi) \end{bmatrix}, \begin{bmatrix} E_z(\rho,-\Phi) \\ E_\rho(\rho,-\Phi) \end{bmatrix} = -Z_b \begin{bmatrix} H_\rho(\rho,-\Phi) \\ -H_z(\rho,-\Phi) \end{bmatrix} \tag{5.2.1}$$

The source is represented by a plane wave at skew incidence

$$E_z^i = E_o e^{j\tau_o\rho\cos(\varphi-\varphi_o)} e^{-j\alpha_o z}, \quad H_z^i = H_o e^{j\tau_o\rho\cos(\varphi-\varphi_o)} e^{-j\alpha_o z}$$

$$E_\rho^i = \frac{1}{j\tau_o^2}\left[\alpha_o \frac{\partial E_z^i}{\partial\rho} + \frac{kZ_o}{\rho}\frac{\partial H_z^i}{\partial\varphi}\right], \quad H_\rho^i = \frac{1}{j\tau_o^2}\left[\alpha_0 \frac{\partial H_z^i}{\partial\rho} - \frac{k}{Z_o\rho}\frac{\partial E_z^i}{\partial\varphi}\right] \tag{5.2.2}$$

where $\alpha_o = k\cos\beta$ and $\tau_o = k\sin\beta$ with $\vartheta = \beta$ skewness zenithal incident angle and $\varphi = \varphi_o$ azimuthal incident angle (Figure 5.2.1).

In particular, we obtain the GWHEs (5.2.3) by taking into account the generalized Wiener-Hopf functional equations of angular regions 1 and 2 (Figure 3.3.2) given in (3.3.57)–(3.3.60) and applying the impedance boundary conditions (5.2.1) (Daniele 2003b; Daniele and Lombardi 2005, 2006).

The system of four generalized Wiener-Hopf equations (GWHEs) in normal form is

$$G(\eta)X_+(\eta) = X_-(m) \tag{5.2.3}$$

with unknown spectral field components

$$X_+(\eta) = \begin{vmatrix} V_{z+}(\eta,0) \\ V_{\rho+}(\eta,0) \\ Z_o I_{z+}(\eta,0) \\ Z_o I_{\rho+}(\eta,0) \end{vmatrix}, \quad X_-(m) = \begin{vmatrix} Z_o I_{\rho+}(-m,\Phi) \\ -Z_o I_{z+}(-m,\Phi) \\ -Z_o I_{\rho+}(-m,-\Phi) \\ Z_o I_{z+}(-m,-\Phi) \end{vmatrix} \tag{5.2.4}$$

$$V_{z+}(\eta,\varphi) = \int_0^\infty E_z(\rho,\varphi) e^{j\eta\rho} d\rho, \; I_{z+}(\eta,\varphi) = \int_0^\infty H_z(\rho,\varphi) e^{j\eta\rho} d\rho \tag{5.2.5}$$

$$V_{\rho+}(\eta,\varphi) = \int_0^\infty E_\rho(\rho,\varphi) e^{j\eta\rho} d\rho, \; I_{\rho+}(\eta,\varphi) = \int_0^\infty H_\rho(\rho,\varphi) e^{j\eta\rho} d\rho \tag{5.2.6}$$

and the GWH kernel

$$G(\eta) = D^{-1}(\eta) S(\eta) \tag{5.2.7}$$

$$S(\eta) = \begin{vmatrix} \xi(\eta) & 0 & -\frac{\alpha_o \eta}{k} & -\frac{\tau_o^2}{k} \\ \frac{\alpha_o \eta}{k} & \frac{\tau_o^2}{k} & \xi(\eta) & 0 \\ \xi(\eta) & 0 & \frac{\alpha_o \eta}{k} & \frac{\tau_o^2}{k} \\ -\frac{\alpha_o \eta}{k} & -\frac{\tau_o^2}{k} & \xi(\eta) & 0 \end{vmatrix},$$

$$D(\eta) = \begin{vmatrix} -n(\eta) z_{11}^a - \frac{\tau_o^2}{k} & -n(\eta) z_{12}^a - \frac{m(\eta)\alpha_o}{k} & 0 & 0 \\ -\frac{m(\eta) z_{11}^a \alpha_o}{k} + \frac{z_{21}^a \tau_o^2}{k} & n(\eta) - \frac{m(\eta) z_{12}^a \alpha_o}{k} + \frac{z_{22}^a \tau_o^2}{k} & 0 & 0 \\ 0 & 0 & -n(\eta) z_{11}^b - \frac{\tau_o^2}{k} & -n(\eta) z_{12}^b - \frac{m(\eta)\alpha_o}{k} \\ 0 & 0 & \frac{m(\eta) z_{11}^b \alpha_o}{k} - \frac{z_{21}^b \tau_o^2}{k} & -n(\eta) + \frac{m(\eta) z_{12}^b \alpha_o}{k} - \frac{z_{22}^b \tau_o^2}{k} \end{vmatrix} \tag{5.2.8}$$

where $m(\eta) = -\eta\cos\Phi + \xi(\eta)\sin\Phi$, $n(\eta) = -\eta\sin\Phi - \xi(\eta)\cos\Phi$ and $\xi(\eta) = \sqrt{\tau_o^2 - \eta^2}$ is the spectral propagation constant along y.

With reference to Figure 3.3.1, the Laplace transforms (5.2.4) are defined as $F_+(\eta,\varphi = 0) = \int_0^\infty f_+(\rho,\varphi = 0) e^{+j\eta\rho} d\rho$ with $\rho = x$ and $F_+(-m,\varphi = \pm\Phi) = \int_0^\infty f_-(\rho,\varphi = \pm\Phi) e^{-jm\rho} d\rho$ with $\rho = v$.

Since $\xi(\eta)$ is a multivalued function, we assume the case with $\xi(0)=\tau_o$ as the proper sheet of the η plane (we recall the vanishing losses assumption in k) and we select the contours originating from the branch points $\eta=\pm\tau_o$ with $\mathrm{Im}[\xi(\eta)]=0$ as branch lines (see section 1.13).

Note that (5.2.3) are generalized equations since plus and minus unknowns ($X_+(\eta)$, $X_-(m)$) are defined into different complex planes η and m that are related together $m=m(\eta)$.

By using the mapping presented in section 3.4

$$\begin{aligned} \eta &= \eta(\bar{\eta}) = -\tau_o \cos\left[\frac{\Phi}{\pi}\left[\arccos\left[-\frac{\bar{\eta}}{\tau_o}\right]\right]\right], \\ \bar{\eta} &= \bar{\eta}(\eta) = -\tau_o \cos\left[\frac{\pi}{\Phi}\left[\arccos\left[-\frac{\eta}{\tau_o}\right]\right]\right] \end{aligned} \tag{5.2.9}$$

we obtain $X_+(\eta(\bar{\eta}))=\bar{X}_+(\bar{\eta})$ and $X_-(m(\bar{\eta}))=\bar{X}_-(\bar{\eta})$ that are respectively plus and minus functions of the new plane $\bar{\eta}$. Details on the properties of the mappings and functions in the new complex plane $\bar{\eta}$ are provided in Chapter 3.

The direct substitution of (5.2.9) into (5.2.3) yields the system of classical Wiener-Hopf equations (CWHEs)

$$\bar{G}(\bar{\eta})\bar{X}_+(\bar{\eta})=\bar{X}_-(\bar{\eta}) \tag{5.2.10}$$

with the matrix WH kernel $\bar{G}(\bar{\eta})=G(\eta(\bar{\eta}))$.

The kernel $\bar{G}(\bar{\eta})$ contains multivalued functions depending on $\bar{\eta}$; thus, also in this plane $\bar{\eta}$, we need to define proper/improper sheets accordingly.

Due to the mapping (5.2.9) that contains the multivalued function

$$\arccos\left(-\frac{\bar{\eta}}{\tau_o}\right)=-j\log\frac{\bar{\eta}+j\sqrt{\tau_o^2-\bar{\eta}^2}}{\tau_o}=-j\log\frac{\bar{\eta}+j\kappa(\bar{\eta})}{\tau_o} \tag{5.2.11}$$

branch points and lines originate from $\kappa(\bar{\eta}) = \sqrt{\tau_o^2 - \bar{\eta}^2}$. We note that $\xi(\bar{\eta}) = \xi(\eta(\bar{\eta})) \neq \kappa(\bar{\eta})$ and we need to assume appropriate proper/improper sheets of the $\bar{\eta}$ plane according to the $\kappa(\bar{\eta})$ function.

We assume the case $\kappa(0) = \tau_o$ as the proper sheet and the contours originating from the branch points $\bar{\eta} = \pm\tau_o$ with $\mathrm{Im}[\kappa(\bar{\eta})] = 0$ as branch lines (for details, see section 3.4).

Note that the definitions of proper and improper sheets are different in the two complex planes η and $\bar{\eta}$.

5.3. Fredholm factorization solution in the $\bar{\eta}$ plane of GWHEs

Taking into consideration the Fredholm factorization method of section 1.9 applied to the CWHEs (5.2.10), we can eliminate the minus vector function $\bar{X}_-(\bar{\eta})$. It yields the Fredholm integral equation of the second kind (FIE)

$$\bar{G}(\bar{\eta})\bar{X}_+(\bar{\eta}) + \frac{1}{2\pi j}\int_{-\infty}^{\infty} \frac{\left[\bar{G}(\bar{\eta}') - \bar{G}(\bar{\eta})\right]\bar{X}_+(\bar{\eta})}{\bar{\eta}' - \bar{\eta}} d\bar{\eta}' = \sum_i \frac{R_{\bar{X}_-}^{n.s.}(\bar{\eta}_i)}{\bar{\eta} - \bar{\eta}_i} + \bar{G}(\bar{\eta})\sum_i \frac{R_{\bar{X}_+}^{n.s.}(\bar{\eta}_i)}{\bar{\eta} - \bar{\eta}_i} \tag{5.3.1}$$

for $\mathrm{Im}[\bar{\eta}] = 0$, $R_{\bar{X}_+}^{n.s.}(\bar{\eta}_i)$ and $R_{\bar{X}_-}^{n.s.}(\bar{\eta}_i)$ are the residues at the poles $\bar{\eta}_i$ of the non-standard parts of $\bar{X}_+(\bar{\eta})$ and $\bar{X}_-(\bar{\eta})$, which correspond to the non-standard part of the respective geometrical optics (GO) components $\bar{X}_+^{GO}(\bar{\eta})$ and $\bar{X}_-^{GO}(\bar{\eta})$ with poles $\bar{\eta}_i$. A brief justification that (5.3.1) is an FIE is made in Appendix 5.A.

We recall that the non-standard parts of the plus and minus functions are related to the poles located respectively in $\mathrm{Im}[\bar{\eta}] > 0$ and $\mathrm{Im}[\bar{\eta}] < 0$, and in (5.3.1), we need to consider only the ones located in the proper sheet of the $\bar{\eta}$ plane due to the application of Fredholm factorization method in this sheet.

Let us consider an incident plane wave impinge on a wedge with incident angle $0 < \varphi_o < \Phi$ (Figure 5.2.1).

From geometrical optics considerations (Snell's law), potential multiple scattered waves are generated from the first reflection of face a with incoming (impinging) angles $\varphi_{n+1} = -\varphi_n + 2\Phi$ and from the first reflection of face b with incoming (impinging) angles $\varphi_{n+1} = -\varphi_n - 2\Phi$. We note that multiple reflections are only possible for $\Phi < \pi / 2$ and increase their number by decreasing Φ .

By exploiting the definition of $\bar{w}$ ($\bar{\eta} = -\tau_o \cos \bar{w}$), as given in section 3.7, we note that only the incident wave pole of the Laplace spectra lies in the proper sheet of the $\bar{\eta}$ plane although all reflected waves (together with the incident one) yield the same point $\bar{\eta}_n = -\tau_o \cos\frac{\pi}{\Phi}\varphi_n = -\tau_o \cos\frac{\pi}{\Phi}\varphi_o = \bar{\eta}_o$ although located in different sheets of the complex plane $\bar{\eta}$.

For this reason, the non-standard part of $\bar{X}_+(\bar{\eta})$ is related only to the incident wave and is present only for $0 < \varphi_o < \Phi / 2$ as $\text{Im}[\bar{\eta}_o] = \text{Im}[-\tau_o \cos(\pi\varphi_o / \Phi)] > 0$.

The Laplace transform (5.2.4) of (5.2.2) for $\varphi = 0$ yields

$$X_+^i(\eta) = \frac{T_o}{\eta + \tau_o \cos \varphi_o} \tag{5.3.2}$$

with

$$T_o = \begin{vmatrix} jE_o \\ j\dfrac{\alpha_o \cos \varphi_o E_o + kZ_o \sin \varphi_o H_o}{\tau_o} \\ jZ_o H_o \\ j\dfrac{\alpha_o Z_o \cos \varphi_o H_o - k \sin \varphi_o E_o}{\tau_o} \end{vmatrix} \tag{5.3.3}$$

It follows the evaluation of the residue in the $\bar{\eta}$ plane from that obtained in the η plane

$$\bar{T}_o = T_o \left.\frac{d\bar{\eta}}{d\eta}\right|_{w=-\varphi_o} = T_o \frac{\pi}{\Phi} \frac{\sin\frac{\pi}{\Phi}\varphi_o}{\sin \varphi_o} \tag{5.3.4}$$

Thus, the non-standard part of $\bar{X}_+(\bar{\eta})$ is

$$R_{\bar{X}_+}^{n.s.}(\bar{\eta}_o) = \bar{T}_o u(\frac{\Phi}{2} - \varphi_o) \tag{5.3.5}$$

where u(x) is the Heaviside step function.

If $\varphi_o > \Phi/2$, $\bar{X}_{+}^{GO}(\bar{\eta})$ is standard while $\bar{X}_{-}^{GO}(\bar{\eta})$ is non-standard, then we need to consider all the non-standard GO components with poles located in the proper sheet of the $\bar{\eta}$ plane (see, for example, equations (58) of Daniele and Lombardi (2006) for reflection coefficients on anisotropic impedance wedge at skew incidence for obtuse angle Φ).

Alternatively, since we know the properties of the GO component $\bar{X}_{+}^{GO}(\bar{\eta})$ of plus function $\bar{X}_{+}(\bar{\eta})$ in the proper sheet (i.e. only the incident wave pole lies in the proper sheet; thus, in the proper sheet, we have only (5.3.2); see the comment in section 3.7), by using (5.3.2), we immediately estimate the residue $R_{\bar{X}_{-}}^{n.s.}$ at $\bar{\eta}_o = -\tau_o \cos\frac{\pi}{\Phi}\varphi_o$ in the form

$$R_{\bar{X}_{-}}^{n.s.}(\bar{\eta}_o) = \bar{G}(\bar{\eta}_o) R_{\bar{X}_{+}}^{s.}(\bar{\eta}_o) = \bar{G}(\bar{\eta}_o)\bar{T}_o u(\varphi_o - \frac{\Phi}{2}) \tag{5.3.6}$$

Substantially, the knowledge of $\bar{G}(\bar{\eta}_o)$ avoids the need of a direct evaluation of the reflection coefficients and related components to estimate the non-standard part of the minus function $\bar{X}_{-}(\bar{\eta})$ in the proper sheet that can be cumbersome to analyze. We note that while (5.3.5) depends only on the incident wave, (5.3.6) also contains reflected ones.

The source term $\bar{F}_o(\bar{\eta})$ of the FIE in (5.3.1) in the proper sheet becomes

$$\bar{F}_o(\bar{\eta}) = \sum_i \frac{R_{\bar{X}_{-}}^{n.s.}(\bar{\eta}_i)}{\bar{\eta}-\bar{\eta}_i} + \bar{G}(\bar{\eta}) \sum_i \frac{R_{\bar{X}_{+}}^{s.}(\bar{\eta}_i)}{\bar{\eta}-\bar{\eta}_i} = \frac{\bar{G}(\bar{\eta}_o)\bar{T}_o}{\bar{\eta}-\bar{\eta}_o} u(\varphi_o - \frac{\Phi}{2}) + \frac{\bar{G}(\bar{\eta})\bar{T}_o}{\bar{\eta}-\bar{\eta}_o} u(\frac{\Phi}{2} - \varphi_o) \tag{5.3.7}$$

The discretization of (5.3.1) with (5.3.7) together with contour warping yields approximate representations of $\bar{X}_{+}(\bar{\eta})$, and the application of (5.2.10) yields $\bar{X}_{-}(\bar{\eta})$.

The selection of a contour B to enhance convergence is extensively discussed in Chapter 1. For example, a practical and appropriate line is $B \equiv B_\theta$

$$B_\theta : \bar{\eta}' = \lambda_\theta(y) = e^{j\theta} y, \quad -\infty < y < \infty \tag{5.3.8}$$

with $0 < \theta < \pi/2$ that makes the branch points $\bar{\eta} = \pm\tau_o$ and other singularities far from the integration line. In fact, the region Ω of deformation between B_θ and the real axis of the $\bar{\eta}$ plane shows no singularities of the integrand (both $\bar{G}(\bar{\eta}'), \bar{X}_+(\bar{\eta}')$); therefore, we obtain

$$\bar{G}(\bar{\eta})\bar{X}_+(\bar{\eta})+\frac{1}{2\pi j}\int_{B_\theta} m_{\bar{\eta}}(\bar{\eta},\bar{\eta}')\bar{X}_+(\bar{\eta}')d\bar{\eta}' = \bar{F}_o(\bar{\eta}), \quad \bar{\eta}\in B_\theta \tag{5.3.9}$$

with integral kernel

$$m_{\bar{\eta}}(\bar{\eta},\bar{\eta}') = If[\bar{\eta}=\bar{\eta}', \bar{G}'(\bar{\eta}), \frac{\bar{G}(\bar{\eta}')-\bar{G}(\bar{\eta})}{\bar{\eta}'-\bar{\eta}}] \tag{5.3.10}$$

5.4. Fredholm factorization solution in the $\bar{w}$ plane of GWHEs

In this section, we adapt the procedure given in section 1.15 to obtain a Fredholm integral equation in the $\bar{w}$ plane of the problem defined in (5.3.1) with (5.3.7) in the $\bar{\eta}$ plane.

The properties of the mapping $\bar{\eta}=-\tau_o\cos\bar{w}$ are given in section 3.7, and the definition of proper sheet of the complex planes depends on the multivalued function $\kappa(\bar{\eta})=\sqrt{\tau_o^2-\bar{\eta}^2}$. We assume the case with $\kappa(0)=\tau_o$ as proper sheet and we select the contours originating from the branch points $\bar{\eta}=\pm\tau_o$ with $\mathrm{Im}[\kappa(\bar{\eta})]=0$ as branch lines (see section 3.4).

To obtain the Fredholm integral equation in the $\bar{w}$ plane, first, we warp the real axis of the $\bar{\eta}$ plane into the straight line $\bar{\lambda}$: $\bar{\eta}' = j\tau_o y/|\tau_o|$, $-\infty< y<\infty$ (see Figure 5.4.1). In general, in wedge scattering problems, there are no singularities of $\bar{G}(\bar{\eta})$ and $\bar{X}_+(\bar{\eta})$ in the region between the real axis and $\bar{\lambda}$ (in the first and third quadrants). In (5.3.1), we obtain that

$$\frac{1}{2\pi j}\int_{-\infty}^{\infty}\frac{\left[\bar{G}(\bar{\eta}')-\bar{G}(\bar{\eta})\right]\bar{X}_+(\bar{\eta})}{\bar{\eta}'-\bar{\eta}}d\bar{\eta}' = \frac{1}{2\pi j}\int_{\bar{\lambda}}^{\infty}\frac{\left[\bar{G}(\bar{\eta}')-\bar{G}(\bar{\eta})\right]\bar{X}_+(\bar{\eta})}{\bar{\eta}'-\bar{\eta}}d\bar{\eta}' \tag{5.4.1}$$

Finally, we obtain from (5.3.1) with (5.3.7) the FIE in the $\bar{w}$ plane using the mappings $\bar{\eta}=-\tau_o\cos\bar{w}$, $\bar{\eta}' = -\tau_o\cos\bar{w}'$

$$\tilde{G}(\bar{w})\tilde{X}_+(\bar{w})+\frac{1}{2\pi j}\int_{\bar{\lambda}_{\bar{w}}}\frac{\left[\tilde{G}(\bar{w}')-\tilde{G}(\bar{w})\right]\tilde{X}_+(\bar{w}')}{-\tau_o\cos\bar{w}'+\tau_o\cos\bar{w}}\tau_o\sin\bar{w}'d\bar{w}' = \tilde{F}_o(\bar{w}) \tag{5.4.2}$$

with $\bar{\eta}_o = -\tau_o \cos \bar{w}_o$,

$$\tilde{F}_o(\bar{w}) = \frac{\bar{G}(\bar{\eta}_o)\bar{T}_o u(\varphi_o - \frac{\Phi}{2}) + \tilde{G}(\bar{w})\bar{T}_o u(\frac{\Phi}{2} - \varphi_o)}{-\tau_o \cos \bar{w} + \tau_o \cos \bar{w}_o} \qquad (5.4.3)$$

where $\bar{\lambda}_{\bar{w}}$ are the images in the $\bar{w}$ plane of the line $\bar{\lambda}$ (with opposite versus) of the $\bar{\eta}$ plane (see Figure 5.4.1):

$$\bar{\lambda}_{\bar{w}} : \bar{w}' = \bar{\lambda}_{\bar{w}}(u) = -\frac{\pi}{2} + ju, \quad -\infty < u < \infty \qquad (5.4.4)$$

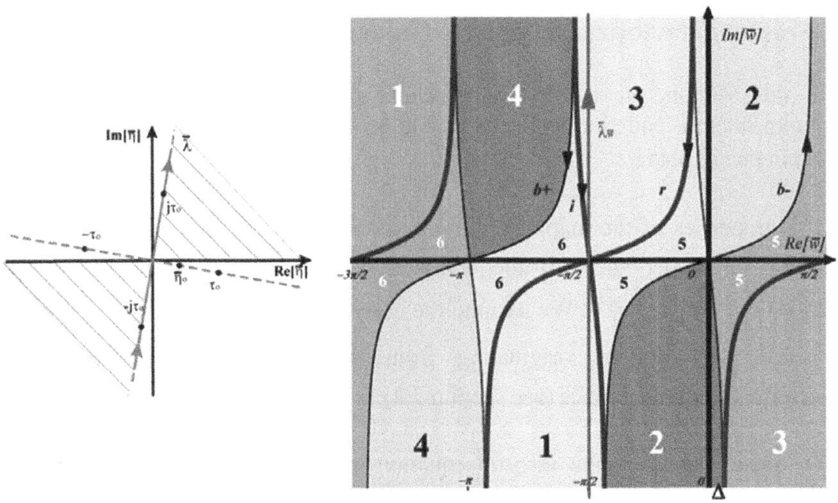

Figure 5.4.1. *Left: contour deformation of integration line from the real axis of the $\bar{\eta}$ plane to $\bar{\lambda}$. Right: image of the integration path $\bar{\lambda}$ in the $\bar{w}$ plane and the image of the proper and improper sheets of the $\bar{\eta}$ plane onto the $\bar{w}$ plane while considering as branch lines the ones with $\mathrm{Im}[\kappa(\bar{\eta})] = 0$. See the caption of Figure 3.4.2 for further details. We define principal proper sheet of the $\bar{\eta}$ plane/$\bar{w}$ plane, the one containing $-\pi < \bar{w} < 0$. For a color version of this figure, see www.iste.co.uk/daniele/scattering2.zip*

We apply $\bar{w}' = -\pi/2 + ju$ ($u \in \mathbb{R}$) to (5.4.2) and we set the observation point on the integration line $\bar{w} = -\pi/2 + jt$ ($t \in \mathbb{R}$); it yields

$$H(t)Y(t) + \frac{1}{2\pi j}\int_{-\infty}^{+\infty} m_t(t,u)Y(u)du = N_o(t), \quad t \in \mathbb{R} \qquad (5.4.5)$$

where

$$m_t(t,u)=\begin{cases}\lim\limits_{u\to t}\dfrac{[H(u)-H(t)]}{-\sinh u+\sinh t}\cosh u, & u=t\\ \dfrac{[H(u)-H(t)]}{-\sinh u+\sinh t}\cosh u, & u\neq t\end{cases} \tag{5.4.6}$$

with $\hat{G}(w)=H(t),\hat{G}(w')=H(u)$, $\tilde{X}_+(\overline{w})=Y(t),\tilde{X}_+(\overline{w}')=Y(u)$, $\tilde{F}_o(\overline{w})=N_o(t)$.

The discretization of (5.4.5) yields approximate representations of $\tilde{X}_+(\overline{w})$, and using inverse mapping (3.4.17) and (5.2.10), we obtain the solution in terms of $\overline{X}_+(\overline{\eta})$ and $\overline{X}_-(\overline{\eta})$.

5.5. Approximate solution of FIEs derived from GWHEs

As stated in section 5.3 and Appendix 5.A, (5.3.9) in the $\overline{\eta}$ plane and the corresponding (5.4.5) in the $\overline{w}$ plane are Fredholm integral equations of the second kind, as derived from (5.3.1). This property allows fast convergence of the integral equations once approximated.

In order to approximate both (5.3.9) and (5.4.5), we use the techniques proposed in Kantorovich and Krylov (1964).

In particular, a careful study of the properties of the integral kernels $m_{\overline{\eta}}(t,u)$ of (5.3.9) and $m_t(t,u)$ of (5.4.5) along the selected integration lines respectively (5.3.8) and $\overline{w}'=-\pi/2+ju$ shows their fast convergence. For this reason, we can apply the simple discretization procedure known as the piecewise constant collocation method (sample-and-hold quadrature method) that truncates the integration interval and discretizes the unknowns with a step parameter.

Applying the method to (5.3.9), we obtain

$$\overline{G}(\lambda_\theta(hn)\overline{X}_+(\lambda_\theta(hn))+\frac{h}{2\pi j}\sum_{i=-A/h}^{A/h} m_{\overline{\eta}}(\lambda_\theta(hn),\lambda_\theta(hi))\overline{X}_+(\lambda_\theta(hi))\lambda_\theta'(hi)=\overline{F}_o(\lambda_\theta(hn)) \tag{5.5.1}$$

where A and h are respectively the truncation and the step parameter such that $A/h\in\mathbb{N}$, and λ_θ is defined in (5.3.8). The selection of A and h is a compromise between computational cost and accuracy that can be further improved diminishing h and increasing A paying attention to the conditioning of the linear system that the discretization generates.

We define the matrices M and D of order 2A/h+1 respectively with non-null elements

$$M_{ni} = \frac{h}{2\pi j} m_{\bar{\eta}}(\lambda_\theta(hn), \lambda_\theta(hi)), \quad n \in [-A/h, A/h], i \in [-A/h, A/h] \tag{5.5.2}$$

$$D_{nn} = \bar{G}(\lambda_\theta(hn), \quad n \in [-A/h, A/h] \tag{5.5.3}$$

and the vector s with elements

$$s_n = \bar{F}_o(\lambda_\theta(hn)), \quad n \in [-A/h, A/h] \tag{5.5.4}$$

the solution f of the linear system

$$[D + M]f = s \tag{5.5.5}$$

provides the samples of $\bar{X}_{a+}(\bar{\eta}) \approx \bar{X}_+(\bar{\eta})$

$$f_n = \bar{X}_{a+}(\lambda_\theta(hn)), \quad n \in [-A/h, A/h] \tag{5.5.6}$$

It yields the approximate solution:

$$\bar{X}_{a+}(\bar{\eta}) = -\frac{[\bar{G}(\bar{\eta})]^{-1} h}{2\pi j} \sum_{i=-A/h}^{A/h} m_{\bar{\eta}}(\bar{\eta}, \lambda_\theta(hi)) \bar{X}_{a+}(\lambda_\theta(hi)) \lambda_\theta'(hi) + [\bar{G}(\bar{\eta})]^{-1} \bar{F}_o(\bar{\eta}) \tag{5.5.7}$$

We note that a local loss of precision in (5.5.7) is for the presence of additional spurious singularities due to the approximation of the integral with a finite sum (Daniele and Zich 2014). We note that (5.5.7) does not guarantee the absence of the singularities of $[\bar{G}(\bar{\eta})]^{-1}$ located $\text{Im}[\bar{\eta}] > 0$ as in the closed form solution and in the exact analytical expression. These singularities are offending for the plus function $\bar{X}_+(\bar{\eta})$, and they do not appear in the exact solution. However, the contribution of these singularities is due to the discretization procedure and their effect is decreasing by increasing the numerical accuracy of the discretization of the integral.

Since (5.3.9) holds in the proper sheet of the $\bar{\eta}$ plane, $\bar{X}_{a+}(\bar{\eta})$ also represents an approximation of $\bar{X}_{+}(\bar{\eta})$ that is limited to a portion of the $\bar{\eta}$ plane due to the fact that (5.5.7) is a sectional analytic representation (Gakhov 1966; Daniele *et al.* 2017a) for the presence of kernel singularity lines of the integral kernel $m_{\bar{\eta}}(\bar{\eta},\bar{\eta}')$: $\bar{\eta}' - \bar{\eta} = 0$, $\eta \neq \eta' \in B_\theta$ (5.3.8). In order to obtain an approximation of $\bar{X}_{+}(\bar{\eta})$ for any $\bar{\eta}$, we need to consider improper sheet evaluations and analytic extension of the initial validity strip of $\bar{X}_{a+}(\bar{\eta})$ as delimited by the singularity lines. In this case, the determination of singularity lines is cumbersome although related to the integration line. This item will be discussed more easily in the $\bar{w}$ plane.

The application of piecewise constant collocation method to (5.4.5) in the $\bar{w}$ plane along (5.4.4) is very similar. We experience that the convergence of the integral equation in the $\bar{w}$ plane is faster than in the $\bar{\eta}$ plane along any contour B_θ (5.3.8) even if tuning the parameter θ. We note that the kernel $m_t(t,u)$ (5.4.6) contains trigonometric functions of $\bar{w}$ that are transformed into hyperbolic trigonometric functions of t, yielding fast converge in $m_t(t,u)$ (graphical representation of the absolute value of $m_t(t,u)$ is similar to a 2D Gaussian point spread function).

Applying the discretization procedure to (5.4.5), we obtain

$$H(hr)Y(hr) + \frac{h}{2\pi j}\sum_{i=-A/h}^{A/h} m_t(hr,hi)Y(hi) = N_o(hr),$$
$$r \in [-A/h, A/h], i \in [-A/h, A/h] \qquad (5.5.8)$$

where A and h are respectively the truncation and the step parameter such that $A/h \in \mathbb{N}$, different from the ones selected for (5.5.1). The comment about the selection of A and h is the same as for (5.5.1). Moreover, in this case, we define a linear system in the matrix form of order 2A/h+1:

$$M_{ri} = \frac{h}{2\pi j} m_t(hr,hi), \quad r \in [-A/h, A/h], i \in [-A/h, A/h] \qquad (5.5.9)$$

$$D_{rr} = H(hr), \quad r \in [-A/h, A/h] \qquad (5.5.10)$$

and the vector s with elements

$$s_r = N_o(hr), \quad r \in [-A/h, A/h] \qquad (5.5.11)$$

The solution f of the linear system

$$[D+M]f = s \tag{5.5.12}$$

provides samples of $Y_a(t) \approx Y(t)$

$$f_n = Y_a(hr), \quad r \in [-A/h, A/h] \tag{5.5.13}$$

It yields the approximate solution:

$$Y_a(t) = -\frac{[H(t)]^{-1} h}{2\pi j} \sum_{i=-A/h}^{A/h} m_t(t, hi) Y(hi) + [H(t)]^{-1} N_o(hr) \tag{5.5.14}$$

Mapping back from t to $\bar{w}$, we obtain

$$\begin{aligned} \tilde{X}_{a+}(\bar{w}) &= -\frac{[\tilde{G}(\bar{w})]^{-1} h}{2\pi j} \sum_{i=-A/h}^{A/h} m_t(-j(\bar{w}+\frac{\pi}{2}), hi) Y_a(hi) + [\tilde{G}(\bar{w})]^{-1} \tilde{F}_o(\bar{w}) \\ \tilde{X}_{a+}(\bar{w}) &= -\frac{[\tilde{G}(\bar{w})]^{-1} h}{2\pi j} \sum_{i=-A/h}^{A/h} \left[\frac{\left(H(hi) - \tilde{G}(\bar{w})\right) Y_a(hi)}{-j\tau_o \sinh(hi) + \tau_o \cos\bar{w}} \tau_o j \cosh(hi) \right] \\ &\quad + [\tilde{G}(\bar{w})]^{-1} \tilde{F}_o(\bar{w}) \end{aligned} \tag{5.5.15}$$

Mapping back $\bar{w} \to \bar{\eta}$ (section 3.7), it yields the approximate representation $\bar{X}_{a+}(\bar{\eta})$, and we can repeat the same comments for analytic continuation already done for solution (5.5.7).

However, if we maintain the representation in the $\bar{w}$ plane, we have the advantage that improper evaluation of multivalued functions is performed via multi-sheets $\bar{w}$ instead via different determinations (e.g. consider $\kappa(\bar{\eta}) = \sqrt{\tau_o^2 - \bar{\eta}^2}$ that yields $\kappa(\bar{w}) = -k \sin \bar{w}$).

Moreover maintaining the representation in the $\bar{w}$ plane, $\tilde{X}_{a+}(\bar{w})$ represents an approximation of $\tilde{X}_+(\bar{w})$ that is limited to a portion of the $\bar{w}$ plane due to the fact that (5.5.15) is a sectional analytic representation (Gakhov 1966; Daniele *et al.* 2017a) for the presence of kernel singularity lines of the integral kernel $m_t(-j(\bar{w}+\frac{\pi}{2}), -j(\bar{w}'+\frac{\pi}{2}))$: $\cos\bar{w} = \cos\bar{w}'$, $\bar{w} \neq \bar{w}' \in \lambda_{\bar{w}}$ (5.4.4).

In order to obtain an approximation of $\tilde{X}_+(\overline{w})$ for any $\overline{w}$, we need analytic extension of the initial validity strip of $\tilde{X}_{a+}(\overline{w})$ as delimited by the singularity lines.

In this case, the determination of singularity lines is much easier than in the $\overline{\eta}$ plane since they are vertical lines parallel to the integration line $\overline{\lambda}_{\overline{w}}$ and located in $\cos\overline{w}=\cos\overline{w}'$, $\overline{w}\neq\overline{w}'\in\overline{\lambda}_{\overline{w}}$ (the closest lines to $\overline{\lambda}_{\overline{w}}$ are located in $\overline{w}'=-\frac{3}{2}\pi+ju, \overline{w}'=+\frac{1}{2}\pi+ju \quad -\infty<u<\infty$). The implementation of the analytic continuation technique in w will be the topic of the next section. We recall that w and $\overline{w}$ are related together by $\overline{w}=\frac{\pi}{\Phi}w$; therefore, transposing properties from each other is straightforward.

We complete the discussion with an alternative solution method for both (5.3.9) and (5.4.5) that consists of recursively applying the approximate representations $\overline{X}_{a+}(\overline{\eta})$ (5.5.7) and $Y_a(t)$ (5.5.14) with starting solution the ones obtained directly from (5.5.7) and (5.5.14) by omitting the integrals, i.e.

$$\overline{X}^{(0)}_{a+}(\overline{\eta})=[\overline{G}(\overline{\eta})]^{-1}\overline{F}_o(\overline{\eta}),\ Y^{(0)}_a(t)-[H(t)]^{-1}N_o(hr) \tag{5.5.16}$$

We recall in particular that the integrals (5.3.9) and (5.4.5) do not contain GO singularities and that the structural singularities of the solutions are inside the spectrum of $[\overline{G}(\overline{\eta})]^{-1}$.

This procedure of successive approximations is discussed and validated in Kantorovich and Krylov (1964) for Fredholm integral equations of the second kind.

5.6. Analytic continuation of approximate solutions of GWHEs

Since both (5.3.9) and (5.4.5) are derived for the principal proper sheet described in section 5.4 and Figure 5.4.1, and due to the presence of kernel singularity lines, the approximate representations of the spectrum (5.5.7) and (5.5.15) are valid only in a portion of the $\overline{\eta}$ plane/ $\overline{w}$ plane, as discussed in section 5.5.

For this reason, to obtain a complete solution in the $\overline{\eta}$ plane/ $\overline{w}$ plane, we need analytic extension. This extension is required for far-field computation via asymptotics with the knowledge of spectral field components in the complex plane in $\bar{\eta}$ and $\bar{w}$ (proper and improper sheets). In fact, asymptotics estimation of the diffracted field in angular regions is performed more easily in the w plane, as anticipated in sections 3.12 and 2.8.

First of all, we note that the principal proper sheet contains the segment of the real axis of the $\bar{w}$ plane delimited by $-\pi < \bar{w} < 0$ (Figure 5.4.1), which corresponds in the w plane to $-\Phi < w < 0$.

In this section, we make a generalization of the procedure given in section 3.9 for a PEC wedge at normal incidence to the general impenetrable wedge scattering problem.

As discussed in section 5.5, here, we want to extend an initial validity strip of the approximate solution whose determination is cumbersome while dealing with (5.5.7) in the $\bar{\eta}$ plane, easier when dealing with (5.5.15) in the $\bar{w}$ plane (the strip is delimited by the kernel singularity lines $\bar{w}' = -\frac{3}{2}\pi + ju, \bar{w}' = +\frac{1}{2}\pi + ju - \infty < u < \infty$). We recall that w and $\bar{w}$ are related together by $\bar{w} = \frac{\pi}{\Phi} w$; therefore, transposing properties from each other is straightforward.

Starting from (5.2.3), by applying the mapping $\eta = -k\cos w$, we obtain, according to section 3.4

$$\hat{G}(w)\hat{X}_+(w) = \hat{Y}_+(w+\Phi) \tag{5.6.1}$$

where $\bar{X}_+(\bar{\eta}) = \tilde{X}_+(\bar{w}) = \hat{X}_+(w)$, $\bar{G}(\bar{\eta}) = \tilde{G}(\bar{w}) = \hat{G}(w)$ and $\bar{X}_-(\bar{\eta}) = \tilde{X}_-(\bar{w}) = \hat{X}_-(w) = Y_+(-m(\bar{\eta})) = Y_+(-\tau_o \cos\frac{\Phi}{\pi}(\bar{w}+\pi)) = \tilde{Y}_+(\bar{w}+\pi) = \hat{Y}_+(w+\Phi)$.

To explicitly obtain $\hat{G}(w)$ from (5.2.7), we need

$$\xi = -\tau_o \sin w, \quad m = \tau_o \cos(w+\Phi), \quad n = \tau_o \sin(w+\Phi) \tag{5.6.2}$$

Replacing $w+\Phi$ with w in (5.6.1) yields

$$\hat{G}(w-\Phi)\hat{X}_+(w-\Phi) = \hat{Y}_+(w) \tag{5.6.3}$$

Using the evenness of $\hat{Y}_+(w)$, we have

$$\hat{G}(w-\Phi)\hat{X}_+(w-\Phi)=\hat{Y}_+(w)=\hat{Y}_+(-w)=\hat{G}(-w-\Phi)\hat{X}_+(-w-\Phi) \qquad (5.6.4)$$

Finally, replacing w with $w-\Phi$, we obtain

$$\hat{G}(w-2\Phi)\hat{X}_+(w-2\Phi)=\hat{G}(-w)\hat{X}_+(-w) \qquad (5.6.5)$$

and using the evenness of $\hat{X}_{+}(w)$ and inverting the kernel, we have

$$\hat{X}_+(w)=[\hat{G}(-w)]^{-1}\hat{G}(w-2\Phi)\hat{X}_+(w-2\Phi) \qquad (5.6.6)$$

We recall that $\hat{X}_+(w)$ is an even function of w, thus we know $\hat{X}_+(w)$ in a vertical strip of wideness $-\Phi<w<+\Phi$

Therefore, once we know the spectrum $\hat{X}_+(w)$ in a vertical strip of wideness 2Φ, we can obtain by applying recursively (5.6.6) the spectrum $\hat{X}_+(w)$ in any point w of the complex plane.

Therefore, (5.6.6) is useful to analytically extend the approximate spectrum $\hat{X}_{a+}(w)=\tilde{X}_{a+}(\overline{w})$ obtained (5.5.15) in the $\overline{w}$ plane that has initial validity strip delimited by the kernel singularity lines $w=-\frac{3\Phi}{2}+ju, w=+\frac{\Phi}{2}+ju \quad -\infty<u<\infty$.

Starting with $\hat{X}_{a+}(w)$ having an initial vertical strip of validity $-\Phi<\mathrm{Re}[w]<0$, we can implement the recursion

$$\hat{X}_{a+}^{ext}(w)=\begin{cases}\hat{X}_{a+}(w), & -\Phi<w<0\\ \hat{X}_{a+}(-w), & 0<w<+\Phi\\ [\hat{G}(-w)]^{-1}\hat{G}(w-2\Phi)\hat{X}_{a+}^{ext}(w-2\Phi), & w>+\Phi\\ \hat{X}_{a+}^{ext}(-w), & w<-\Phi\end{cases} \qquad (5.6.7)$$

5.7. Far-field computation

In order to estimate the total far field of the scattering problem constituted of an impedance wedge illuminated at skew incidence by a plane wave, we can use either inverse Laplace transform of the spectrum observed for an arbitrary direction φ or equivalently using the Sommerfeld functions. Both the spectra observed for an arbitrary direction φ and the Sommerfeld functions are defined in Chapter 3.

In the following, we omit the $e^{-j\alpha_o z}$ dependence and limit the details for the Ez component while Hz can be derived similarly. With reference to Figure 5.2.1, the other electromagnetic field components of incident (i) and scattered (s) waves can be derived from the Ez and Hz components: $E_z^i = \sin\beta E_\beta^i$, $H_z^i = \frac{1}{Z_o}\sin\beta E_\varphi^i$, $E_\beta^s = -\frac{1}{\sin\beta}E_z^s$, $E_\varphi^s = -\frac{1}{\sin\beta}Z_o H_z^s$

Starting from the inverse Laplace transform of the spectrum $V_{zd}(\eta,\varphi)$ given in terms of axial spectra via (3.10.5), we have

$$E_z(\rho,\varphi) = \frac{1}{2\pi}\int_{B_r} V_{z+}(\eta,\varphi)e^{-j\eta\rho}d\eta = \frac{\tau_o}{2\pi}\int_{\lambda(B_r)} \hat{V}_{zd}(w,\varphi)e^{j\tau_o\rho\cos w}dw \qquad (5.7.1)$$

where B_r is the Bromwich contour (a horizontal line over the singularities of $V_{z+}(\eta,\varphi)$) and $\lambda(B_r)$ is the contour mapped into the w plane.

By deforming the contour $\lambda(B_r)$ into the SDP path defined by

$$w = -\pi + gd(\mathrm{Im}[w]) + j\,\mathrm{Im}[w], \quad gd(x) = \mathrm{sgn}(x)\arccos(1/\cosh(x)) \qquad (5.7.2)$$

with gd being the Guderman function and $w = -\pi$ being the saddle point, we obtain

$$E_z(\rho,\varphi) = -\frac{\tau_o e^{-jk_o\rho}}{2\pi}\int_{\mathrm{SDP}} \hat{V}_{zd}(w,\varphi)e^{\tau_o\rho h(w)}dw - j\tau_o\sum_i \mathrm{Res}\left[\hat{V}_{zd}(w,\varphi)\right]_{w_i(\varphi)} e^{+j\tau_o\rho\cos w_i(\varphi)} \qquad (5.7.3)$$

where $w_i(\varphi) = w_{oi} \pm \varphi$ are the poles of $\hat{V}_{zd}(w,\varphi)$ and w_{oi} are the poles of the axial spectrum $\hat{V}_{zd}(w) = \hat{V}_{zd}(w,\varphi = 0)$.

The far-field components of (5.7.3) are classified into

$$E_z(\rho,\varphi) = E_z^g(\rho,\varphi) + E_z^d(\rho,\varphi) + E_z^s(\rho,\varphi) \qquad (5.7.4)$$

where $E_z^g(\rho,\varphi)$ is the GO field originates from the GO poles (5.7.5), $E_z^s(\rho,\varphi)$ is the surface wave component due to the poles of WH kernel (5.7.5) and $E_z^d(\rho,\varphi)$ is the diffracted component (5.7.6) resulting from the SDP integral:

$$E_z^{g,s}(\rho,\varphi) = -j\tau_o \sum_i \text{Res}\left[\hat{V}_{zd}(w,\varphi)\right]_{w_i(\varphi)} e^{+j\tau_o\rho\cos w_i(\varphi)} \quad (5.7.5)$$

$$E_z^d(\rho,\varphi) = -\frac{\tau_o e^{-jk_o\rho}}{2\pi}\int_{\text{SDP}} \hat{V}_{zd}(w,\varphi)e^{\tau_o\rho h(w)}dw \quad (5.7.6)$$

The propagation of GO waves assumes the form $e^{+j\tau_o\rho\cos(w_i(\varphi))} = e^{j\tau_o\rho\cos(w_{oi}\pm\varphi)} = e^{-j\tau_o\vec{\rho}\cdot\hat{\tau}_{oi}}$ with $\hat{\tau}_{oi} = -\hat{x}\cos w_{oi} \pm \hat{y}\sin w_{oi}$.

We note that the GO components are present when the GO poles $w_i(\varphi)$ are captured by the warping between $\lambda(B_r)$ and the SDP path. By varying the observation angle φ, a GO pole can cross the SDP; in this case, the related contribution vanishes in accordance with the shadow boundary defined in geometrical optics.

Since $w_i(\varphi) = w_{oi} \pm \varphi$, w_{oi} are the poles of the axial spectrum $\hat{V}_{zd}(w) = \hat{V}_{zd}(w,\varphi = 0)$ and $w = -\pi$ is the saddle point, the poles that can generate shadow regions by crossing the saddle point are the ones in $-\pi - \Phi < w < +\pi + \Phi$ as $-\Phi < \varphi < +\Phi$, see also (3.10.5).

Moreover, to compute GO components, we do not need to solve the Fredholm integral equation since the integral term in it is regular for any GO pole.

The diffracted component (5.7.6) can be estimated asymptotically with the saddle point technique, yielding

$$E_z^d(\rho,\varphi) \approx E_z^{gtd}(\rho,\varphi) = E_o \frac{e^{-j(\tau_o\rho + (\pi/4))}}{\sqrt{2\pi\tau_o\rho}} D_E(\varphi,\varphi_o) \quad (5.7.7)$$

with the GTD diffraction coefficient

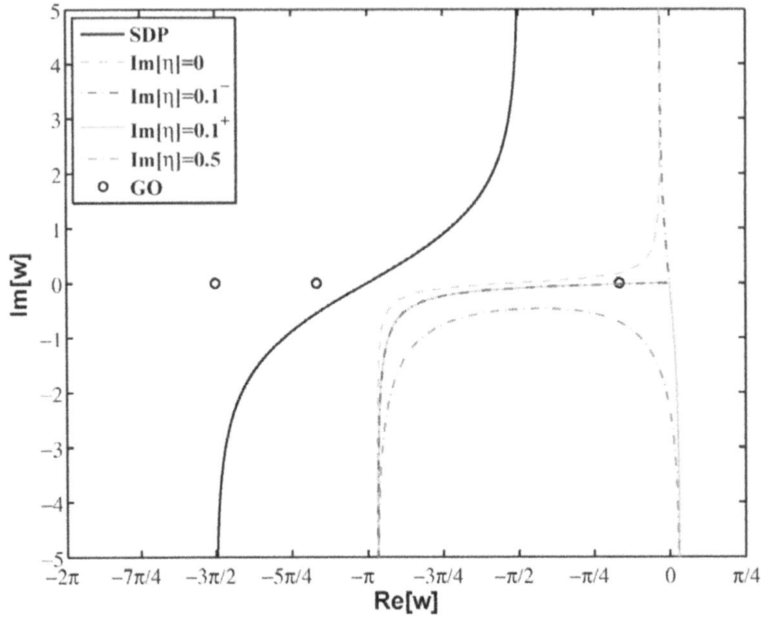

Figure 5.7.1. *Horizontal ($\mathrm{Im}[\eta]=\cos t$) Bromwich contours B_r and SDP contour in the w plane with branch points at $\pm\tau_o$ and $\tau_o=1-0.1j$. The positive $\mathrm{Re}[\eta]=\cos t$ direction in B_r corresponds to the direction of the contours shown in the figure with direction towards $-\pi+\arctan(-\mathrm{Im}[k]/\mathrm{Re}[k])-j\infty=-3.042-j\infty$. The circles represent the GO poles of the axial spectra illustrated for a test case with $\Phi=3\pi/2$, $\varphi_o=\pi/6$: $w_{oi}=-\varphi_o,-2\Phi+\varphi_o,-2\Phi-\varphi_o$ corresponding to the incident wave, reflected wave from face a and reflected wave from face b. For a color version of this figure, see www.iste.co.uk/daniele/scattering2.zip*

$$D_E\left(\varphi,\varphi_o\right)=\frac{-\tau_o\hat{V}_{zd}(-\pi,\varphi)}{jE_o} \tag{5.7.8}$$

Since $-\Phi<\varphi<+\Phi$, due to the expression of $V_{zd}(\eta,\varphi)$ in terms of axial spectra via (3.10.5), it becomes clear that we need to know the axial spectra at least in $-\pi-\Phi<w<+\pi+\Phi$ to compute (5.7.8). For this reason, we need to apply the analytic extension procedure (5.6.6) on the initial approximate spectrum $\hat{X}_{a+}(w)=\tilde{X}_{a+}(\overline{w})$ obtained from (5.5.15) in the $\overline{w}$ plane. From section 5.6, in fact, the initial validity strip is delimited by $w'=-\frac{3}{2}\Phi+ju, w'=+\frac{1}{2}\Phi+ju\quad -\infty<u<\infty$.

The uniform expression of the diffracted field is obtained through the UTD (Kouyoumjian and Pathak 1974), which removes the caustics of GTD for the observation angle φ aligned with the GO shadow boundaries $\varphi_q+\pi$ (φ_q is the incoming direction of the related wave)

$$E_z^{\mathrm{utd}}(\rho,\varphi)=E_o\frac{e^{-j(\tau_o\rho+(\pi/4))}}{\sqrt{2\pi\tau_o\rho}}C_E(\varphi,\varphi_o) \tag{5.7.9}$$

$$C_E(\varphi,\varphi_o)=D_E(\varphi,\varphi_o)+\sum_q\Gamma_q\frac{1-F\left(2\tau_o\rho\cos^2\frac{\varphi-\varphi_q-\pi}{2}\right)}{\cos\frac{\varphi-\varphi_q-\pi}{2}} \tag{5.7.10}$$

and $F(z)$ is the Kouyoumjian–Pathak transition function used and defined in section 2.8 at (2.8.5).

By applying the Sommerfeld functions, we obtain similar results.

In fact, starting from the Sommerfeld integral for Ez (Hz with a similar procedure)

$$E_z(\rho,\varphi)=\frac{1}{2\pi j}\left[\int_C s_E(w+\varphi)e^{+j\tau_o\cos(w)\rho}dw\right] \tag{5.7.11}$$

where C is the Sommerfeld contour shown in section 3.12 and Figure 3.12.2.

For immediate availability, here we define the Sommerfeld function in terms of axial spectra

$$\begin{aligned}s_E(w)=\frac{j}{2}\Bigg[&-\tau_o\sin wV_{z+}(-\tau_o\cos w,0)+\frac{\tau_o^2}{\omega\varepsilon}I_{\rho+}(-\tau_o\cos w,0)\\&-\frac{\alpha_o\tau_o\cos w}{\omega\varepsilon}I_{z+}(-\tau_o\cos w,0)\Bigg]\\s_H(w)=\frac{j}{2}\Bigg[&-\tau_o\sin wI_{z+}(-\tau_o\cos w,0)-\frac{\tau_o^2}{\omega\mu}V_{\rho+}(-\tau_o\cos w,0)\\&+\frac{\alpha_o\tau_o\cos w}{\omega\mu}V_{z+}(-\tau_o\cos w,0)\Bigg]\end{aligned} \tag{5.7.12}$$

As for the inverse Laplace transform, we apply the SDP method to the Sommerfeld integral with the SDP paths crossing the saddle points $w = \pm\pi$ (Figure 3.12.2). It yields $E_z(\rho,\varphi) = E_z^g(\rho,\varphi) + E_z^d(\rho,\varphi) + E_z^s(\rho,\varphi)$ where the GO and surface wave components are given in terms of the poles $w_i(\varphi)$ (singularities of both $s_E(w+\varphi)$ and $\hat{V}_{zd}(w,\varphi)$ by definition).

$$E_z^{g,s}(\rho,\varphi,z) = \sum_i \operatorname{Res}\left[s_E(w+\varphi)\right]_{w=w_i(\varphi)} e^{+j\tau_o \cos w_i(\varphi)\rho} \tag{5.7.13}$$

and where the diffracted component $E_z^d(\rho,\varphi)$ is approximated with GTD using the saddle point method

$$E_z^{gtd}(\rho,\varphi) = E_o \frac{e^{-j(\tau_o\rho+(\pi/4))}}{\sqrt{2\pi\tau_o\rho}} D_E(\varphi,\varphi_o) \tag{5.7.14}$$

The GTD diffraction coefficient is given by

$$D_E(\varphi,\varphi_o) = \frac{s_E(\varphi-\pi) - s_E(\varphi-\pi)}{E_o} \tag{5.7.15}$$

Since (5.7.13) and (5.7.15) are alternative expressions of (5.7.5) and (5.7.7), shadow boundary considerations and uniformization done before still hold.

In order to extract the GO components in practical applications, we recall from section 3.7 that by supposing an incident wave with $0 < \varphi_o < \Phi$, we have the potential sets of multiple reflected waves (from Snell's law) starting from face "a" $\varphi_{n+1} = -\varphi_n + 2\Phi$ and starting from face "b" $\varphi_{n+1} = -\varphi_n - 2\Phi$, yielding the complete set $\varphi_n = -\varphi_o + 2n\Phi$ ($n = 0, \pm1, \pm2, \ldots$). As stated, these waves correspond in the axial spectra to the point $\bar{\eta}_n = -\tau_o \cos\frac{\pi}{\Phi}\varphi_n = -\tau_o \cos\frac{\pi}{\Phi}\varphi_o = \bar{\eta}_o$ (section 5.3) but in different sheets of the complex plane $\bar{\eta}$. In this case, to estimate the GO field from (5.7.5) and (5.7.13), we need the corresponding values of the GO poles in the w plane that are $w_n = \pm\varphi_n$. Alternatively, we can compute GO fields by just considering geometrical optics.

In order to extract the surface wave contributions in practical applications, we note that they are structural singularities, i.e. originating from singularities of the WH kernel (5.2.7) at its inverse; see (5.5.7) and (5.5.15):

$$G(\eta)= D^{-1}(\eta)S(\eta)\,,\ [G(\eta)]^{-1} = S(\eta)^{-1}D(\eta) \tag{5.7.16}$$

Let us work with $D(\eta)$ (5.2.8) written in the m plane, i.e. $D(m)$.

In the general case of anisotropic non-symmetric impedance faces, we have singularities due to the determinants of $S(\eta)$ and $D(m)$. We have

$$\mathrm{Det}[S(\eta)] = -\frac{4\left(\tau_o^2-\eta^2\right)\tau_o^4}{k^2}\,,\ \mathrm{Det}[D(m)] = -\frac{\tau_o^4}{k^4}d_a(m)d_b(m) \tag{5.7.17}$$

$$\begin{aligned} d_{a,b}(m) = \left(n^2+\alpha_o^2\right)z_{11}^{a,b} + kn\left(1+\mathrm{Det}\left[\mathbf{Z}_{a,b}\right]\right) \\ - m\alpha_o\left(z_{12}^{a,b}+z_{21}^{a,b}\right) + \tau_o^2 z_{22}^{a,b} \end{aligned} \tag{5.7.18}$$

Note that while $\mathrm{Det}[S(\eta)]$ shows only the zeros $\pm\tau_o$ that are coincident with the branch points, $d_{a,b}(m)=0$ can yield multiple new zeros.

Considering the w plane, from (5.6.2), the zeros $d_{a,b}(m)=0$ correspond to the zeros of

$$\begin{aligned} \hat{d}_{a,b}(w+\Phi) = \left(\tau_o^2\sin^2(w+\Phi)+\alpha_o^2\right)z_{11}^{a,b} + k\tau_o\sin(w+\Phi)\left(1+\mathrm{Det}\left[\mathbf{Z}_{a,b}\right]\right) \\ -\tau_o\cos(w+\Phi)\alpha_o\left(z_{12}^{a,b}+z_{21}^{a,b}\right)+\tau_o^2 z_{22}^{a,b} \end{aligned} \tag{5.7.19}$$

While estimating (5.7.5) and (5.7.13), the poles satisfy $\hat{d}_{a,b}(\pm(w_i(\varphi)+\varphi)-\Phi)=0$ by applying (5.6.7). In fact, the location of surface wave poles is typically close to the SDP path on Sommerfeld function asymptotic estimation (5.7.12).

In particular, these poles are described by recursion formula (5.6.7). While observing the spectra for $w<-\Phi$, we note the presence of $\hat{G}(-w-2\Phi)$ and thus the singularities of $D^{-1}(m)$, i.e. the singularities of $\hat{d}_{a,b}(w+\Phi)$, need to be considered substituting w with $-w-2\Phi$, thus we need to compute the zeros of

$\hat{d}_{a,b}(-w-\Phi)$. A similar consideration holds for $w>\Phi$ with possible surface waves lying on the other face of the wedge.

5.8. Criteria for the examples

In this section, we illustrate examples of scattering by impenetrable wedge. In particular, with reference to Figure 5.1.2, the examples depend on the following parameters:

1. aperture angle of the wedge in terms of Φ;
2. impedances of the two faces $\mathbf{Z}_a = Z_o \mathbf{z}_a, \mathbf{Z}_b = Z_o \mathbf{z}_b$;
3. propagation constant $k = k' - jk''$ of the medium;
4. kind of illumination by plane wave:
   a. skewness according to the zenithal angle of incidence β,
   b. azimuthal angle of incidence φ_o,
   c. in the case of normal incidence ($\beta = \pi/2$), polarization (Ez or Hz),
   d. intensity of plane wave in terms of z components E_o and H_o;
5. discretization parameters for the FIE: truncation A and step h.

The angles are always assumed in radians.

The solution via the FIE is validated by comparison with other independent solutions or by self-convergence. We propose some interesting practical engineering/physical results in electromagnetic scattering by impenetrable wedge. Without loss of generality, in the following, we will assume $k = 1 - j10^{-8}$. Repetition of a simulation for different practical values of k yields similar performance.

Alternatively, the analysis of the problem for practical values of k' is obtainable by scaling the quantities obtained for $k = 1 - j10^{-8}$. In particular, a different value of k' (e.g. $k' = p$ such that $k_{new} = p(1 - j10^{-8})$ changes the computed spectrum $F_+(\eta)$ into $(k/k_{new})F_+(k\eta/k_{new})$; see section 2.8. This property also shows that the GTD diffraction coefficient (5.7.8) is independent of k.

The parameters of validation are relative errors computed on the spectra and on the estimation of far field.

5.9. Example 1: Symmetric isotropic impedance wedge at normal incidence with Ez polarization

This first simple example is analyzed completely in terms of all the tools presented in this book.

In particular, (5.2.3)–(5.2.8) are considerably simplified.

In fact, for Ez polarization, we have ($H_o = 0$); thus, $E_\rho(\rho,\varphi)=0, H_z(\rho,\varphi)=0$ and the GWHEs (5.2.3) result

$$\begin{cases} G(\eta)X_+(\eta) = X_-(m) \\ G(\eta) = \begin{vmatrix} -\dfrac{\xi}{Z_a(n+n_a)} & \dfrac{n_a}{(n+n_a)} \\ \dfrac{\xi}{Z_a(n+n_a)} & \dfrac{n_a}{(n+n_a)} \end{vmatrix} \\ X_+(\eta) = \begin{vmatrix} V_{z+}(\eta,0) \\ I_{\rho+}(\eta,0) \end{vmatrix}, \quad X_-(m) = \begin{vmatrix} I_{\rho+}(-m,\Phi) \\ I_{\rho+}(-m,-\Phi) \end{vmatrix} \end{cases} \tag{5.9.1}$$

where $Z_a = Z_o z_a, n_a = \omega\mu / Z_a = k\sin\vartheta_a$.

After some mathematical manipulations on (5.9.1), i.e. sum and subtraction of the two equations, we obtain

$$\begin{cases} \dfrac{\xi}{n+n_a} V_{z+}(\eta,0) = \dfrac{Z_a}{2}\left[I_{\rho+}(-m,-\Phi) - I_{\rho+}(-m,\Phi)\right] \\ \dfrac{n_a}{n+n_a} I_{\rho+}(\eta,0) = \dfrac{1}{2}\left[I_{\rho+}(-m,-\Phi) + I_{\rho+}(-m,\Phi)\right] \end{cases} \tag{5.9.2}$$

We perform the analysis using three strategies:

1) using the classical factorization of the scalar kernel with logarithmic factorization on (5.9.2) (see sections 3.7 and 1.7);

2) using an approximate computation of the factorization of the scalar kernel on (5.9.2) with the Fredholm factorization method in $\bar{\eta}$ rewritten in $\bar{w}$ (see section 1.17), noting that factorization is defined up to a constant;

3) direct semi-analytical solution of the FIE derived from matrix equation (5.9.1) in the $\bar{w}$ plane (see section 5.5).

The first method is based on the scalar factorization in $\bar{\eta}$ with the support of the w plane of

$$\bar{g}_1(\bar{\eta}) = \frac{\xi}{n+n_a} = -\frac{\sin w}{\sin(w+\Phi)+\sin\vartheta_a}, \quad \bar{g}_2(\bar{\eta}) = \frac{n_a}{n+n_a} = \frac{\sin\vartheta_a}{\sin(w+\Phi)+\sin\vartheta_a} \tag{5.9.3}$$

as

$$\hat{g}_{1+}(w) = \frac{\sin w}{d_\Phi(w)\sin\frac{\pi}{\Phi}w}, \quad \hat{g}_{2+}(w) = \frac{1}{d_\Phi(w)\cos\frac{\pi}{2\Phi}w} \tag{5.9.4}$$

in terms of

$$d_\Phi(w) = \left[\frac{n+n_a}{n}\right]_+ = \exp\left[\frac{1}{\pi}\int_0^\infty \arctan\left(\frac{1}{z_a\sinh\left(\frac{\Phi u}{\pi}\right)}\right)\frac{\sinh u}{\cosh u+\cos w}du\right] \tag{5.9.5}$$

In the following, $d_\Phi(w)$ is evaluated at the machine precision level.

Therefore, we obtain the classical solution (3.7.9) and Daniele (2003b) in the w plane

$$\begin{cases} \hat{V}_{z+}(w,0) = -j\dfrac{\pi}{\Phi}\dfrac{\sin\frac{\pi}{\Phi}\varphi_o}{k\sin\varphi_o}\dfrac{\hat{g}_{1+}(-\varphi_o)}{\hat{g}_{1+}(w)}\dfrac{E_o}{\cos\left(\frac{\pi}{\Phi}w\right)-\cos\left(\frac{\pi}{\Phi}\varphi_o\right)} \\ \hat{I}_{\rho+}(w,0) = j\dfrac{\pi}{\Phi}\dfrac{\sin\frac{\pi}{\Phi}\varphi_o}{kZ_o}\dfrac{\hat{g}_{2+}(-\varphi_o)}{\hat{g}_{2+}(w)}\dfrac{E_o}{\cos\left(\frac{\pi}{\Phi}w\right)-\cos\left(\frac{\pi}{\Phi}\varphi_o\right)} \end{cases} \tag{5.9.6}$$

To illustrate the numerical results, we select as parameters of the problem (see section 5.8 for glossary): $\Phi = 3\pi/4$, $z_a = 0.5$, $k = 1 - j10^{-8} m^{-1}$, $\beta = \pi/2$, $\varphi_o = 7\pi/12$, $E_o = 1V/m, H_o = 0A/m$ and A, h such that $A = 20, h = 0.5$ unless otherwise specified.

These parameters of the problem are similar to those reported in Senior and Volakis (1995, p. 122); thus, we have a further independent solution based on the SM method to compare the results in terms of the total far field.

First of all, we make a comparison on the kernel factorization obtained from the first and second strategies. Figure 5.9.1 shows the absolute values of $\hat{g}_{1+}(w)$ and $\hat{g}_{2+}(w)$ for $-\Phi < w < 0$ obtained with strategy 1 and machine precision evaluation of $d_\Phi(w)$. In the same figure, we have computed the relative error in log10 scale of the results obtained with strategy 2 for A = 20, h = 0.5 and fictitious pole $\eta_{of} = -k\cos(\varphi_{of})$ ($\varphi_{of} = 2\pi/3$) with respect to strategy 1. From the figure, we note a good agreement of the results that yields at least 4 digits of precision without optimizing A, h and φ_{of}.

In Figure 5.9.2, we present the numerical results in terms of axial spectra ($\hat{V}_{z+}(w) = \hat{V}_{z+}(w,0)$, $\hat{I}_{\rho+}(w) = \hat{I}_{\rho+}(w,0)$) obtained with three strategies. In particular, we have plotted the absolute value obtained with strategy 1 and machine precision evaluation of $d_\Phi(w)$. In the same figure we have computed the relative errors with respect to strategy 1 of the solutions obtained with strategy 2 ($\varphi_{of} = 2\pi/3$) and strategy 3. Both strategies 2 and 3 are implemented for A = 20, h = 0.5. Moreover, in the figure, we note a good agreement of the results that yields at least 4 digits of precision without optimizing A, h and φ_{of}.

In Figure 5.9.3, we validate strategy 3 with respect to strategy 1 by studying convergence in terms of integration parameters A and h. The figure shows, in particular, the relative error in log10 scale of the results in terms of $\hat{V}_{z+}(w) = \hat{V}_{z+}(w,0)$ obtained using strategy 3 with respect to strategy 1. We have selected in the tests A = 20,40,60; h = 1,0.5,0.1.

We note that in this problem, we saturate the precision of $\hat{V}_{z+}(w) = \hat{V}_{z+}(w,0)$ for A = 40, h = 0.5. As stated in the theory, the precision should increase by decreasing h and increasing A, paying attention to the conditioning of the linear system of the Fredholm factorization method in strategy 3 (see section 5.5).

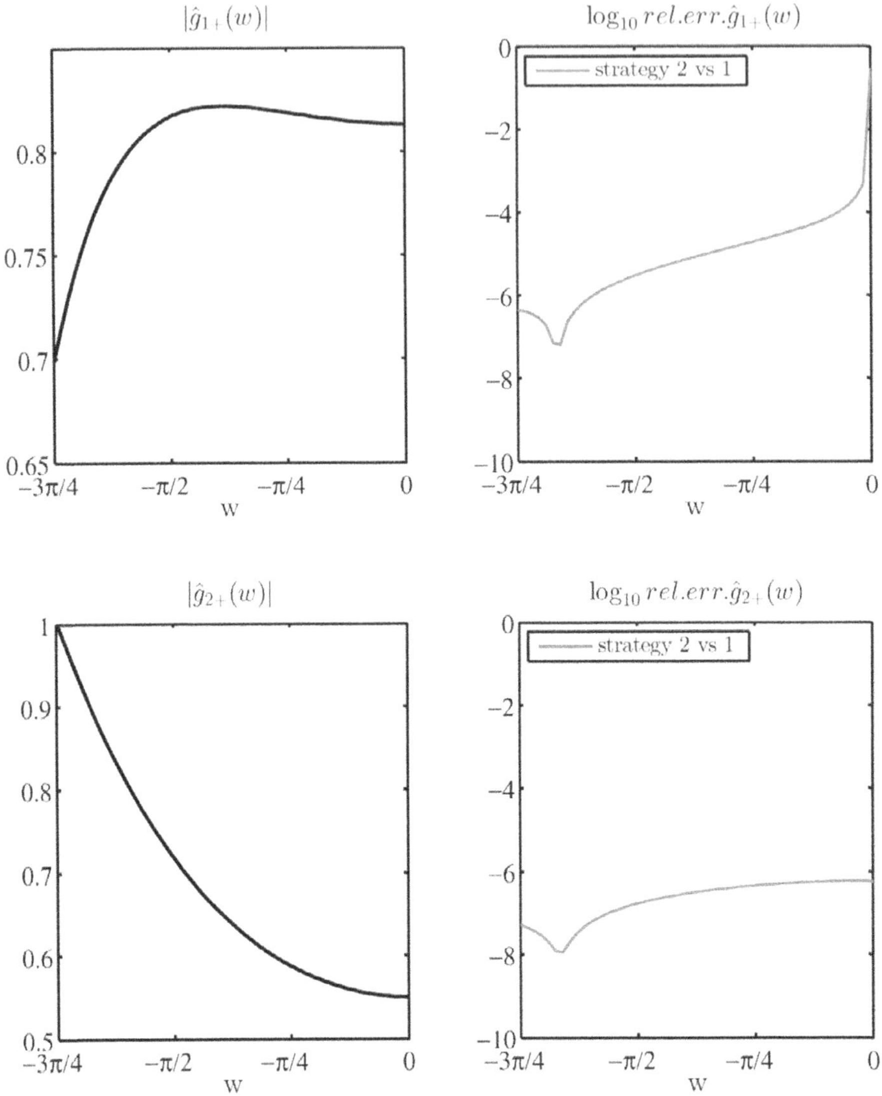

Figure 5.9.1. *Top-left (bottom-left): absolute value of* $\hat{g}_{1+}(w)$ $(\hat{g}_{2+}(w))$ *for* $-\Phi < w < 0$ *obtained with strategy 1 and machine precision evaluation of* d_Φ. *Top-right (bottom-right): relative error in log10 scale of the results obtained with strategy 2 for A = 20, h = 0.5 and fictitious pole* $\eta_{of} = -k\cos(\varphi_{of})$ $(\varphi_{of} = 2\pi/3)$ *with respect to strategy 1*

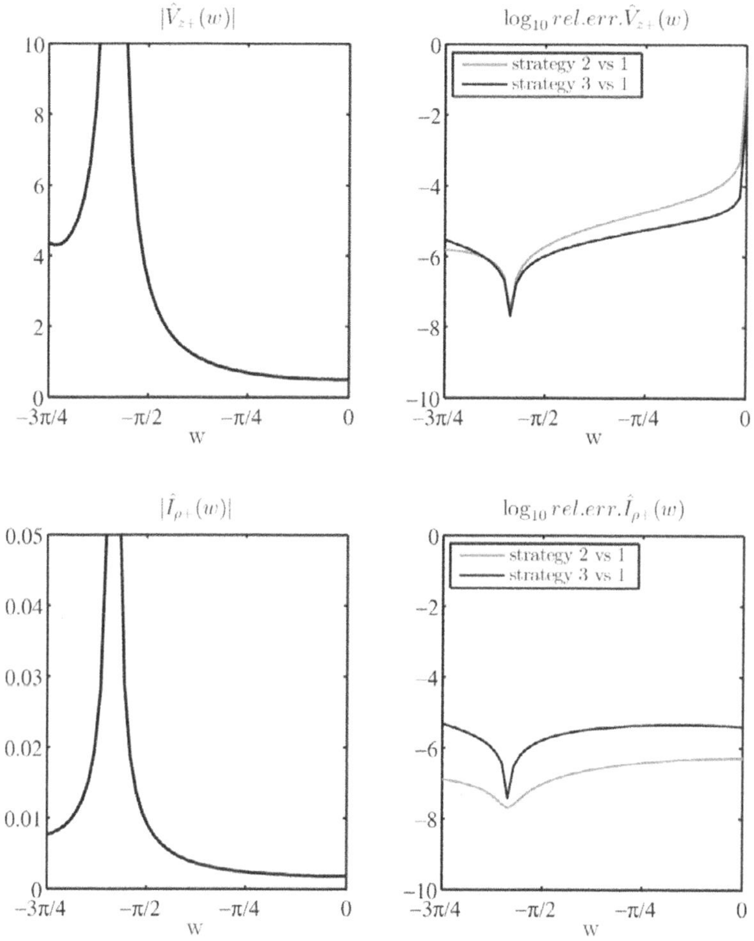

Figure 5.9.2. *Top-left (bottom-left): absolute value of* $\hat{V}_{z+}(w)=\hat{V}_{z+}(w,0)$ $(\hat{I}_{\rho+}(w)=\hat{I}_{\rho+}(w,0))$ *for* $-\Phi<w<0$ *obtained with strategy 1 and machine precision evaluation of* $d_\Phi(w)$. *Top-right (bottom-right): relative error in log10 scale of the results obtained with strategy 2 with pole* $\varphi_{of}=2\pi/3$ *and strategy 3 (for both A = 20 and h = 0.5) with respect to strategy 1*

We recall that the three strategies give approximated representations of $\hat{V}_{z+}(w)=\hat{V}_{z+}(w,0)$ and $\hat{I}_{\rho+}(w)=\hat{I}_{\rho+}(w,0)$ that are valid only in a portion of spectral planes. In particular, Figure 5.9.4 shows the approximations obtained by the three strategies before applying analytic extension via (5.6.7). In the figure, the results are compared to the extended spectra, which are obtained by one of the three strategies after analytic extension (5.6.7).

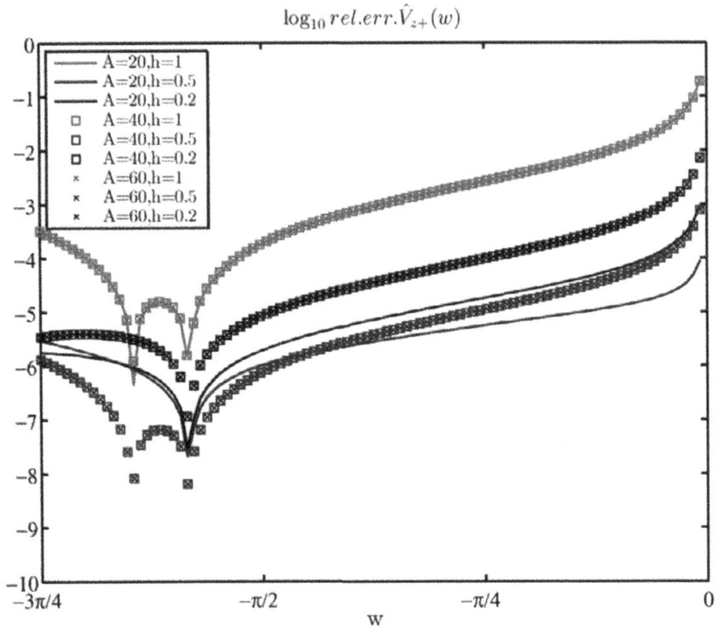

Figure 5.9.3. *Plot of the relative error in log10 scale of the results in terms of* $\hat{V}_{z+}(w)=\hat{V}_{z+}(w,0)$ *obtained with strategy 3 parameterized for A,h with respect to strategy 1. For a color version of this figure, see www.iste.co.uk/daniele/scattering2.zip*

An important observation is that the initial spectrum of strategy 1 covers the proper sheet of $\bar{\eta}$ mapped in the w plane, and it implements the even symmetry of the plus function. The result is that strategy 1 directly gives a correct spectrum in $-\Phi < w < \Phi$.

For what concerns strategies 2 and 3, since both are based on Fredholm factorization implemented in the w domain, we expect that the initial spectra is delimited by the kernel singularity lines, see sections 5.5 and 5.6. In fact, Figure 5.9.4 shows that strategies 2 and 3 have an initial spectrum that overlaps and is correct in $-3\Phi/2 < w < \Phi/2$.

From Figure 5.9.4, we also note that the correct extended spectrum has poles at $w=\pm\varphi_o, \pm(2\Phi-\varphi_o), \pm(2\Phi+\varphi_o)=\pm 7\pi/12, \pm 11\pi/12, \pm 25\pi/12$ (the last two lie outside of the figure) that correspond to the incident wave, reflected wave from face a and reflected wave from face b (the last does not occur in this problem since it is not captured by the inverse Laplace transform).

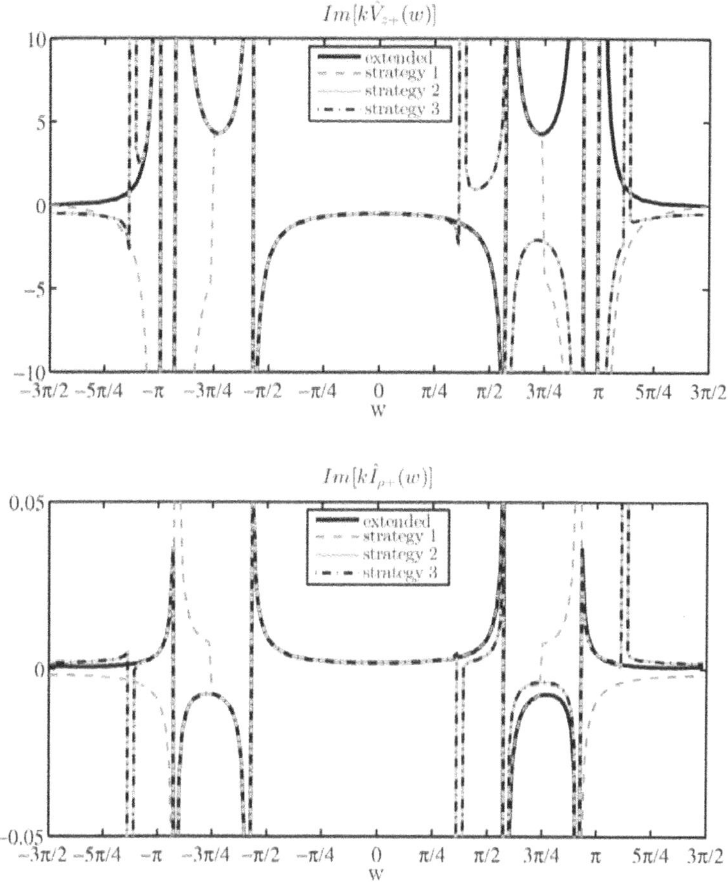

Figure 5.9.4. *Top (bottom): imaginary part of the initial spectrum* $k\hat{V}_{z+}(w) = k\hat{V}_{z+}(w,0)$ ($k\hat{I}_{\rho+}(w) = k\hat{I}_{\rho+}(w,0)$) *for* $-2\Phi < w < 2\Phi$ *obtained with strategies 1, 2 and 3 and plot of the spectrum correctly extended using analytic extension (5.6.7). The initial spectrum of strategy 1 (2 and 3) is correct in* $-\Phi < w < \Phi$ ($-3\Phi/2 < w < \Phi/2$) *with* $\Phi = 3\pi/4$. *For a color version of this figure, see www.iste.co.uk/daniele/scattering2.zip*

Finally, Figure 5.9.5 presents the results in terms of asymptotic evaluation of the far field, as discussed in section 5.7, using the approximate spectral solution obtained using strategy 3 with A = 20, h = 0.5. In particular, the GO field is

$$\begin{aligned} E_z^g = {} & E_o e^{j\tau_o\rho\cos(\varphi-\varphi_o)}\left[u(\pi-|\varphi-\varphi_o|)-u(\varphi-\Phi)\right] \\ & + R_a e^{j\tau_o\rho\cos(\varphi+\varphi_o-2\Phi)}\left[u(\pi-|\varphi+\varphi_o-2\Phi|)-u(\varphi-\Phi)\right] \end{aligned} \tag{5.9.7}$$

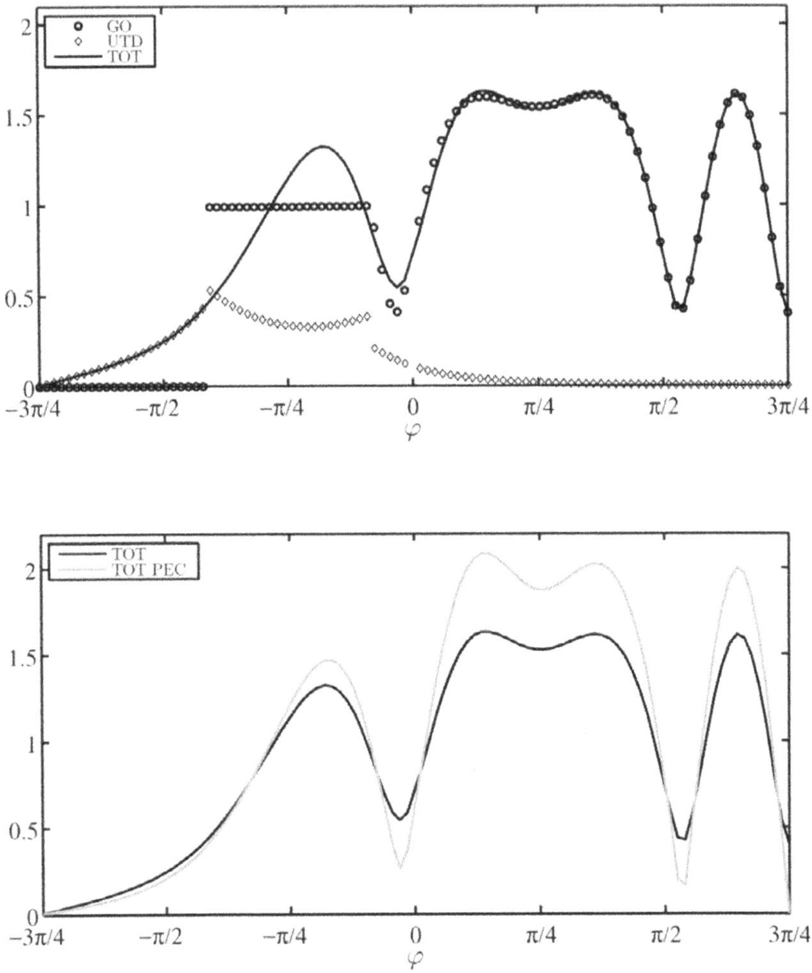

Figure 5.9.5. *Top: total field (solid line), GO field component (circles) and UTD field component (diamond) at* $k\rho = 10$ *from the edge of the wedge for the solution obtained with strategy 3 with A = 20, h = 0.5. Bottom: comparison of the total field obtained for the problem of section 5.9 with the equivalent problem with impedance* $z_a = 0$ *(PEC wedge)*

with reflection coefficient of face a

$$R_a = \frac{E_o(-1 + z_a sin(\Phi - \varphi_o))}{1 + z_a sin(\Phi - \varphi_o)} = -0.6 \quad (5.9.8)$$

For what concerns the UTD component, we evaluate (5.7.9).

The figure also presents the same problem but with different impedances, i.e. $z_a = 0$ of the PEC wedge.

As a last independent validation, we recall that the parameters of the problem are similar to that reported in Senior and Volakis (1995). Figure 5.9.5 overlaps with Figure 4.7 of Senior and Volakis (1995, p. 122).

5.10. Example 2: Non-symmetric isotropic impedance wedge at normal incidence with Hz polarization and surface wave contribution

In this section, we apply the Fredholm factorization method to another well-known test case, solved by using the SM method and reported in Norris and Osipov (1999); also studied with the WH technique in (Daniele and Lombardi 2006). With reference to Figure 5.2.1 and to the parameters of the problem, as described in section 5.8: $\Phi = 7\pi/8$, $z_a = 0.01$, $z_b = \sin(\vartheta_b)$ with $\vartheta_b = 0.01 + j$, $k = 1 - j10^{-8} m^{-1}$, $\beta = \pi/2, \varphi_o = \pi/2$, $E_o = 0V/m, H_o = 1A/m$.

The solution is obtained via the direct semi-analytical solution of the FIE in the $\overline{w}$ plane with A = 20, h = 0.5, derived from matrix equations (5.2.3)–(5.2.8) (see section 5.5).

In particular, we have estimated the total field H_z due to an H-polarized plane wave at a distance $k\rho = 10$ from the edge of the wedge.

Figure 5.10.1 presents the total field and its decomposition into three components: the GO field component, the UTD field component and the surface wave (SW) field component.

Note that the only difference from the test case reported in Norris and Osipov (1999) is $z_a = 0.01$ instead of $z_a = 0$. This choice is made to check that our methodology does not have problems for some critical small values of the parameters. Similar results are obtained using $z_a = 0$.

The figure also presents the same problem but with different impedances, i.e. $z_a = z_b = 0$ of the PEC wedge.

Focusing the attention on the SW wave contribution of face b ($\varphi = -\Phi$) as stated at the end of section 5.7, the related poles are, in particular, described by the zeros of $\hat{d}_b(-w - \Phi)$.

In the non-symmetric isotropic impedance wedge, (5.7.19) simplifies to

$$d_b(m) = n^2 z_b + kn\left(1 + z_b^2\right) + k^2 z_b \tag{5.10.1}$$

The zeros of (5.10.1) are

$$n = -kz_b, -k / z_b \tag{5.10.2}$$

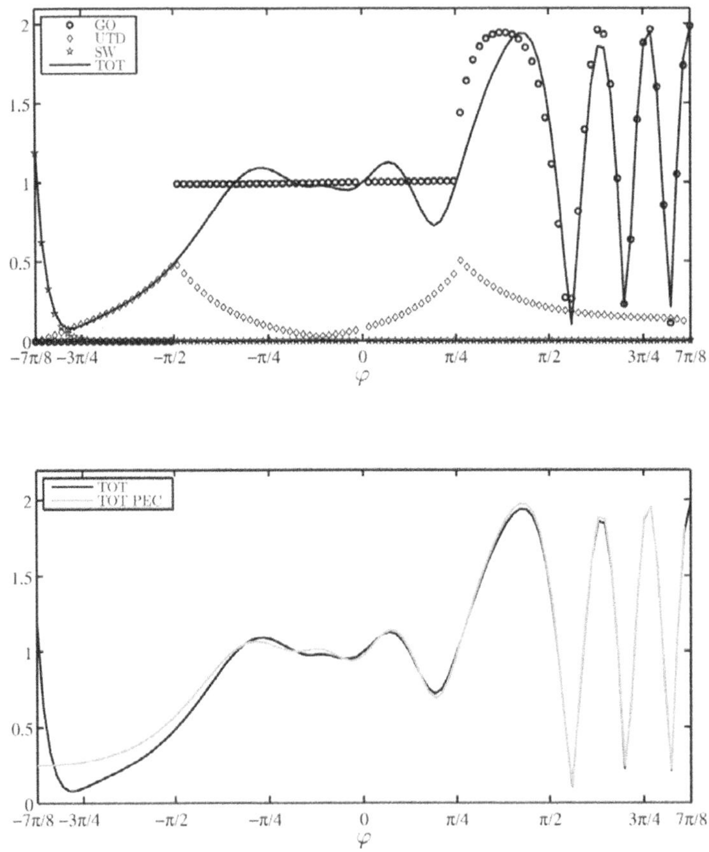

Figure 5.10.1. *Top: total field (solid line), GO field component (circles), UTD field component (diamonds) and SW wave (stars) at* $k\rho = 10$ *obtained with direct application of the Fredholm factorization method with A = 20, h = 0.5. Bottom: comparison of the total field obtained for the problem of section 5.10 with the equivalent problem with impedance* $z_a = z_b = 0$ *(PEC wedge)*

Focusing the attention on $n=-kz_b=-k\sin\vartheta_b$, we have in the w plane

$$\sin(w+\Phi)=-(\sin\vartheta_b) \tag{5.10.3}$$

However, since we need to compute the expression substituting w with $-w-2\Phi$ (section 5.7), we have that the singularities of the axial spectrum are given by

$$\sin(-w-\Phi)=-(\sin\vartheta_b) \tag{5.10.4}$$

Among all the possible solutions, there is

$$w_{swb}=-\pi-\Phi-\vartheta_b \tag{5.10.5}$$

that generates a surface wave on face b. Rigorous considerations on the region where w_{swb} is captured with corresponding uniformization requires the description of the SDP path (see section 5.7) that is omitted here since the decay of the surface wave is fast for $\varphi>-\Phi$.

Finally, to obtain the expression of the SW wave contribution, we apply (5.7.13)

$$E_z^{g,s}(\rho,\varphi,z)=\sum_i \text{Res}\left[\, s_E(w+\varphi)\right]_{w=w_i(\varphi)} e^{+j\tau_o\cos w_i(\varphi)\rho} \tag{5.10.6}$$

with $w_i(\varphi)=w_{swb}-\varphi$. In this problem, we obtain $\left[\, s_E(w+\varphi)\right]_{w=w_i(\varphi)}=$ $0.919-j0.995$ for $w_{swb}=-\dfrac{15\pi}{8}-0.01-j$ (note the location in Figure 3.12.2).

5.11. Example 3: PEC wedge at skew incidence

In this section, we consider the PEC wedge at skew incidence, as a test case, to validate our general vector formulation with a well-known explicit scalar solution for the GTD diffraction coefficient.

The factorization of the kernel (5.2.7) for PEC boundary conditions is reduced to the scalar factorization of the $\xi=\sqrt{\tau_o^2-\eta^2}$ and $n=-\eta\sin\Phi-\xi\cos\Phi$ functions (see section 6.3 and Appendix 4.A of this book and sections 3.3 and 10.16 of Daniele and Zich (2014) for their derivation in the $\bar{w}$ plane). For this reason, we can obtain the classical closed-form expression of GTD diffraction coefficients (5.7.15) in terms of Sommerfeld functions

$$s_E(w) = \pi \frac{E_o \cos\left(\frac{\pi}{2\Phi}\varphi_o\right)}{2\Phi\left[\sin\left(\frac{\pi}{2\Phi}w\right) - \sin\left(\frac{\pi}{2\Phi}\varphi_o\right)\right]} \tag{5.11.1}$$

$$s_H(w) = \pi \frac{H_o \cos\left(\frac{\pi}{2\Phi}w\right)}{2\Phi\left[\sin\left(\frac{\pi}{2\Phi}w\right) - \sin\left(\frac{\pi}{2\Phi}\varphi_o\right)\right]} \tag{5.11.2}$$

The accuracy and the convergence of the approximate solutions obtained with the Fredholm factorization method can be ascertained by the comparison between the exact GTD diffraction coefficients (5.7.15) and the approximate ones obtained with different discretization parameters A and h.

With reference to Figure 5.2.1 and to the parameters of the problem, as described in section 5.8: $\Phi = 7\pi/8$, $z_a = z_b = 0$, $k = 1 - j10^{-8}m^{-1}$, $\beta = \pi/6$, $\varphi_o = 3\pi/4$, $E_o = 1V/m, H_o = 0A/m$.

The exact GTD diffraction coefficient in dB, together with the relative error of the numerical solution in log10 scale, is shown in Figure 5.11.1. The use of the relative error in log 10 scale is a good tool to validate the precision because it measures the digits of precision in the whole interval of the observation angle φ except for the directions where GTD is infinite or almost vanishing.

Peaks of the GTD diffraction coefficients occur for the GO directions: $\varphi = \varphi_o - \pi = -\pi/4$ (incident wave) and for $\varphi = 2\Phi - \varphi_o - \pi = 0$ (face a reflected wave).

The figure shows relative errors less than 10^{-7} for A = 25, h = 0.25 at any observation angle.

This high convergence rate is due to the mathematical properties of the Fredholm integral equation of the second kind in $\overline{w}$ without any regularization technique.

We note that although the analytic expression of the GTD diffraction coefficients is independent of β, the WH vector formulation depends on this parameter.

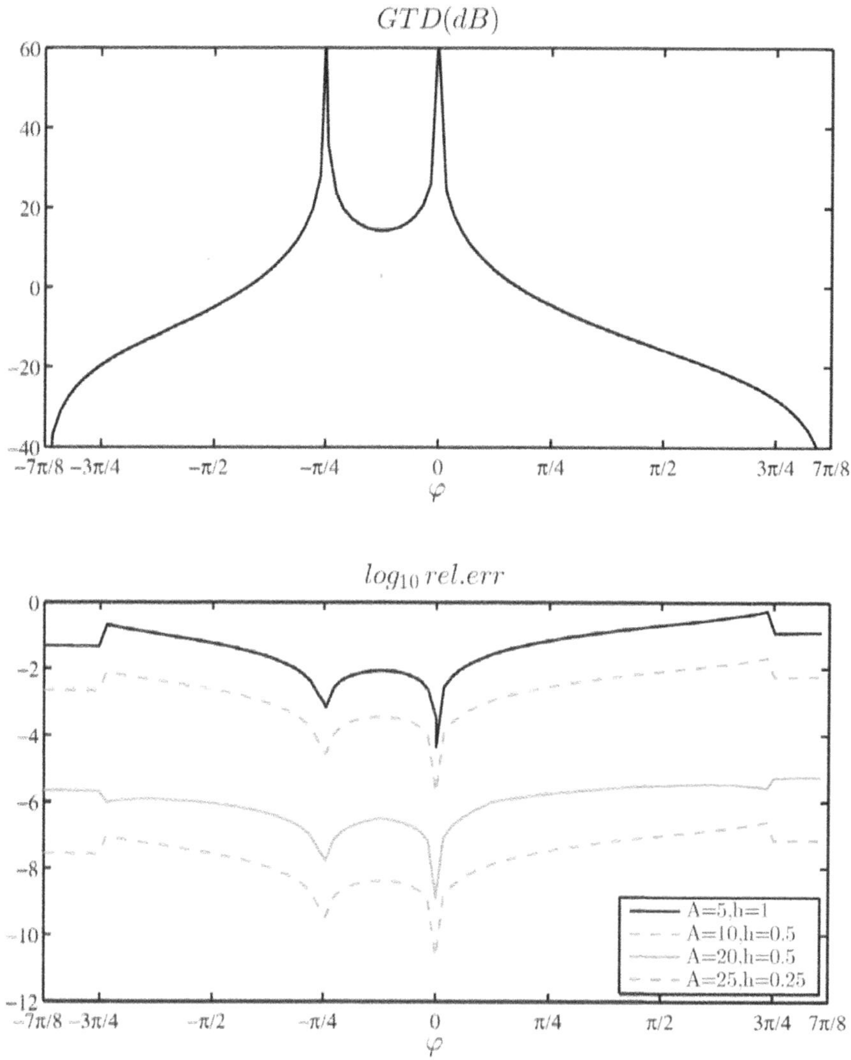

Figure 5.11.1. *Top: the exact GTD diffraction coefficient dB. Bottom: relative error of the evaluation of GTD diffraction coefficient using the Fredholm factorization method with respect to the exact value in log10 scale using different sets of integration parameters for the PEC wedge, described in section 5.11, versus the azimuthal observation angle φ*

5.12. Example 4: Arbitrary impedance half-plane at skew incidence

We take inspiration for the test case of an arbitrary impedance half-plane at skew incidence from that reported in Senior and Volakis (1995, p. 130), in particular generalizing it to non-symmetric anisotropic impedance faces.

With reference to Figure 5.2.1 and the parameters of the problem as described in section 5.8: $\Phi = \pi$, impedances $\mathbf{z}_a, \mathbf{z}_b$ as given below, $k = 1 - j10^{-8} m^{-1}$, $\beta = \pi / 3$, $\varphi_o = 5\pi / 6$, $E_o = 1V / m, H_o = 0A / m$.

The approximate solutions obtained with the Fredholm factorization method are obtained with discretization parameters A = 10 and h = 1.

Figure 5.12.1 shows the GTD co-polar diffraction coefficient in dB for the field component Ez' versus the azimuthal observation angle φ with different sets of $\mathbf{z}_a, \mathbf{z}_b$ from scalar to matrix (the scalars are the same as those used in Senior and Volakis (1995)):

1) isotropic face impedances $z_a = 0,\ z_b = 0$;

2) isotropic face impedances $z_a = 0.25,\ z_b = 4$;

3) isotropic face impedances $z_a = 10,\ z_b = 0$;

4) anisotropic face impedances $\mathbf{z_a} = \begin{bmatrix} 2-j & 1+2j \\ -0.5 & 1-j \end{bmatrix}, \mathbf{z_b} = \begin{bmatrix} 0.5 & 0 \\ 0 & 2.6 \end{bmatrix}$.

Note that for the anisotropic test case we have chosen passive impedance matrices (Zhu and Landstorfer 1997).

Figure 5.12.2 proposes the study of convergence using the set of impedance $z_a = 0.25,\ z_b = 4$. The figure shows the relative error in log10 scale on the GTD diffraction coefficient for the solution obtained with the Fredholm factorization method with different integration parameters A and h. This is a self-convergence test and we take the one with A = 25, h = 0.25 as the reference solution.

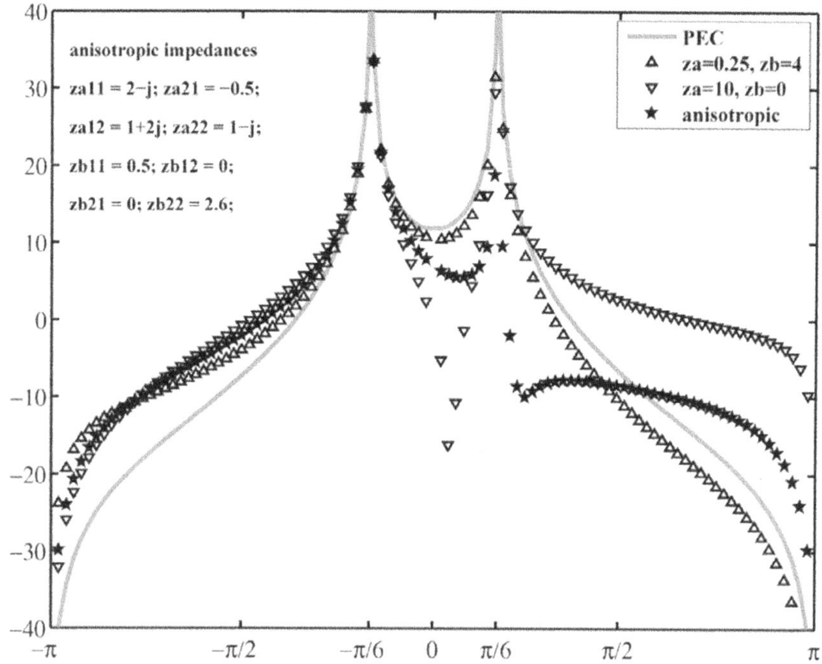

Figure 5.12.1. *GTD co-polar diffraction coefficient in dB versus the azimuthal observation angle φ for different sets of surface impedances for the half-plane problem at skew incidence (figure modified and adapted from Daniele and Lombardi (2006)), using the Fredholm factorization method with A=10, h=1. For a color version of this figure, see www.iste.co.uk/daniele/scattering2.zip*

In Figure 5.12.2, we establish a satisfactory choice of the integration parameters in terms of digit of precision $A > 10, h < 0.5$ taking into account that an excessive A and a small h can yield ill conditioning in the linear system associated with the method.

In particular, for values of A>40 and h<0.25, no improvement of the solution is achieved.

The same numerical performance is observed for the other sets of impedance (isotropic and anisotropic).

As stated previously, this high convergence rate is due to the mathematical properties of the Fredholm integral equation of the second kind in $\overline{w}$ without any regularization technique for the spectral properties of the integral kernel in the t plane ($\bar{w} = -\pi/2 + jt$).

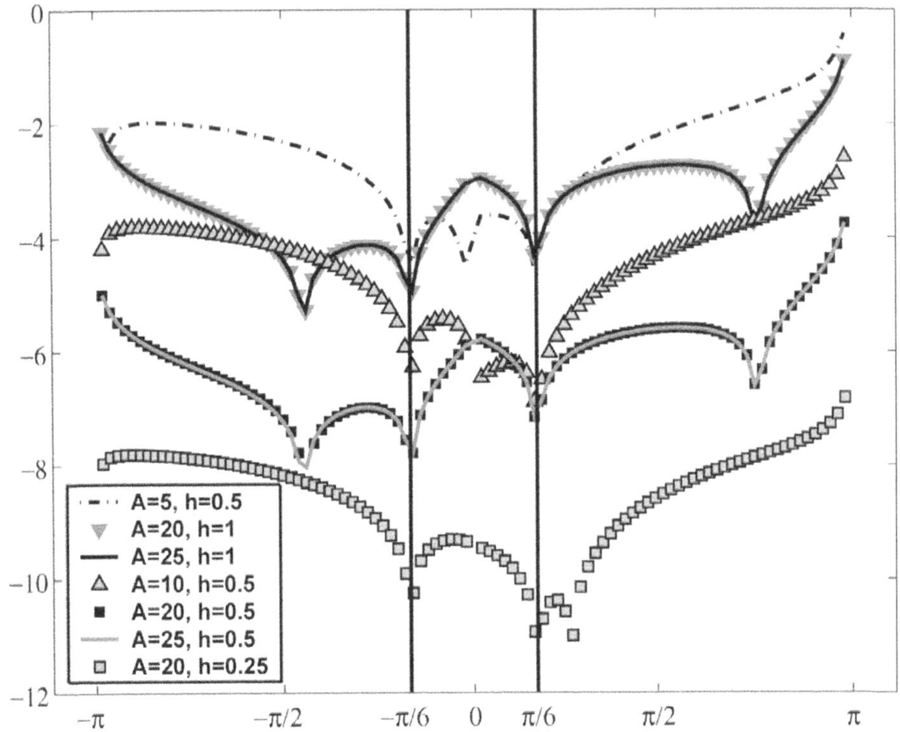

Figure 5.12.2. *The relative error on the evaluation of GTD diffraction coefficient in log10 scale using a different set of integration parameters A, h for the test case with* $z_a = 0.25$, $z_b = 4$ *(© 2006 IEEE[1]). For a color version of this figure, see www.iste.co.uk/daniele/scattering2.zip*

5.13. Example 5: Arbitrary impedance wedge at skew incidence

In this section, we present numerical results for the evaluation of the total field scattered and diffracted by a non-symmetric isotropic impenetrable wedge at skew incidence. We evaluate the total far-field components E_z and $Z_o H_z$.

1 Reprinted with permission from V.G. Daniele and G. Lombardi, Wiener-Hopf solution for impenetrable wedges at skew incidence, *IEEE Trans. Antennas Propag.*, vol. 54, no. 9, pp. 2472–2485, 2006.

With reference to Figure 5.2.1, the parameters of the problem are: $\Phi = 7\pi/8$, $z_a = 0.25, z_b = 4$, $k = 1 - j10^{-8}m^{-1}$, $\beta = \pi/3$, $\varphi_o = \pi/2$, $E_o = 1V/m, H_o = 0A/m$.

The approximate solutions obtained with the Fredholm factorization method are given by discretization parameters A = 10 and h = 1.

Numerical results of the total far field are presented in Figures 5.13.1 and 5.13.2 respectively for the co-polar component E_z and the cross-polar component $Z_o H_z$ specifying the GO field component and the UTD field component.

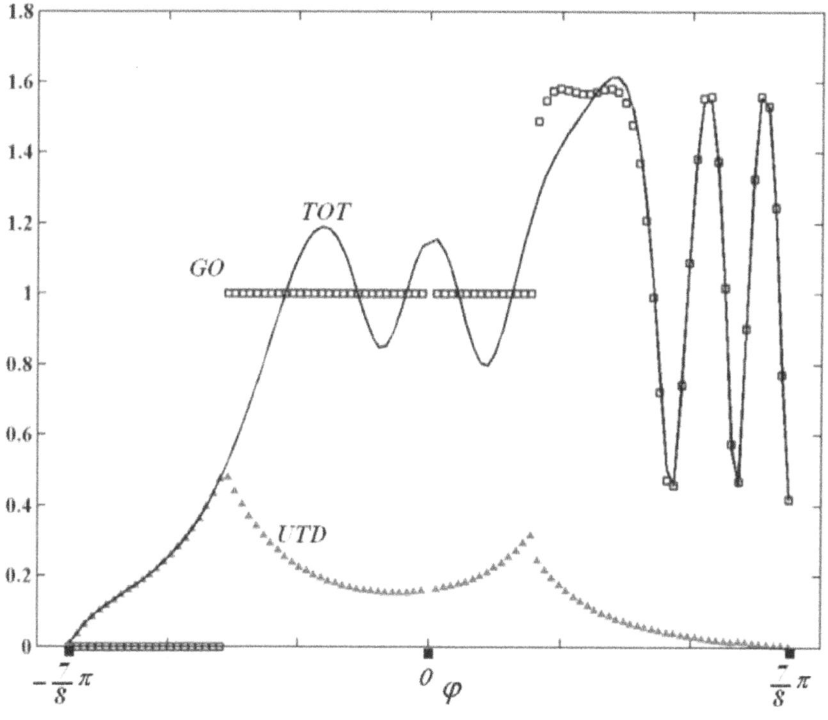

Figure 5.13.1. *Co-polar component E_z of total far field at $k\rho = 10$ for the test case of section 5.13: total field (solid line), GO field component (squares) and UTD field component (triangles) (© 2006 IEEE[2])*

2 Reprinted with permission from V. G. Daniele and G. Lombardi, Wiener-Hopf Solution for Impenetrable Wedges at Skew Incidence, *IEEE Trans. Antennas Propag.*, vol. 54, no. 9, pp. 2472–2485, 2006.

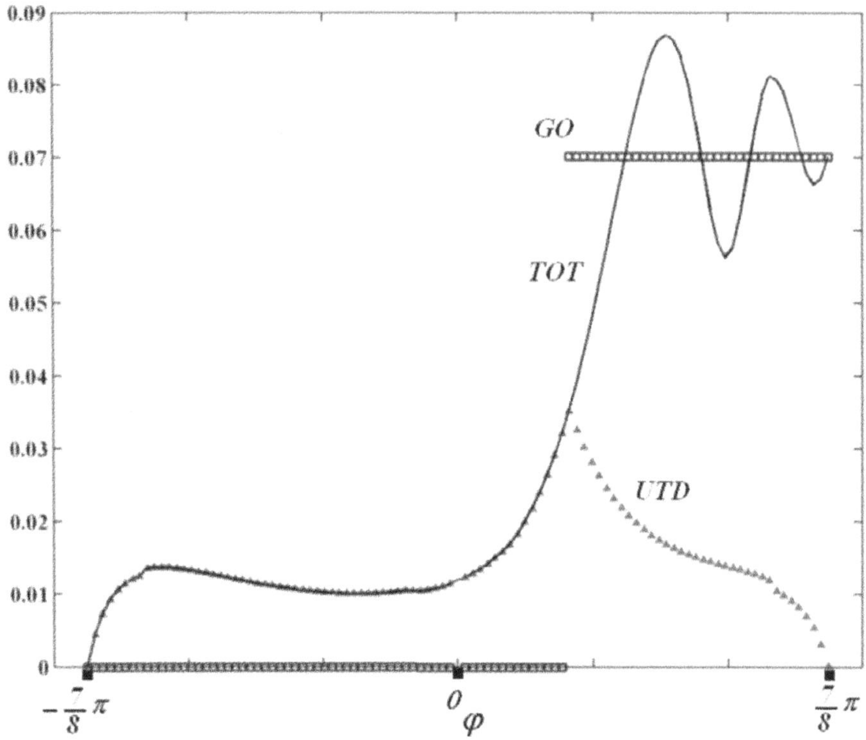

Figure 5.13.2. *Cross-polar component $Z_o H_z$ of total far field at $k\rho = 10$ for the test case of section 5.13: total field (solid line), GO field component (squares) and UTD field component (triangles) (© 2006 IEEE[3])*

5.14. Example 6: Arbitrary impedance concave wedge at skew incidence

This example demonstrates the capability of the method to model a concave wedge, i.e. $\Phi < \pi / 2$, which in general excites numerous reflected waves that might yield non-standard components in the axial spectra. Without loss of generality, we consider $0 < \varphi_o < \Phi$.

3 Reprinted with permission from V. G. Daniele and G. Lombardi, Wiener-Hopf Solution for Impenetrable Wedges at Skew Incidence, *IEEE Trans. Antennas Propag.*, vol. 54, no. 9, pp. 2472–2485, 2006.

From section 5.3 we observe that we have two sets of incoming (impinging) GO wave angles: $\varphi_{n+1} = -\varphi_n + 2\Phi$ when the first reflection is from face a and $\varphi_{n+1} = -\varphi_n - 2\Phi$ when the first reflection is from face b.

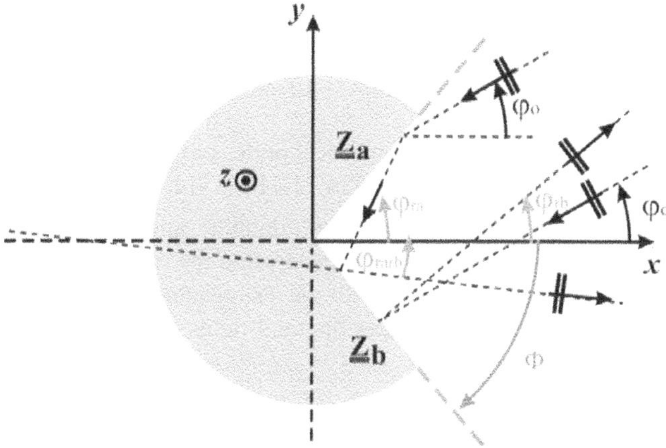

Figure 5.14.1. *Impendence concave wedge with multiple reflections. For a color version of this figure, see www.iste.co.uk/daniele/scattering2.zip*

However, the source of the integral equations (5.3.1), (5.4.2) and (5.4.5) can also be computed in this case by knowing just the spectra of the incident field on the plus function, as described in section 5.3, skipping all the details on the non-standard components of plus and minus functions in the proper sheet.

This property is remarkable, and we restate it in this context, where we have two homogeneous angular regions of the same arbitrary aperture angle.

We anticipate that in more general complex scattering problems (e.g. regions of different materials and shapes, such as two different angular regions or an angular region coupled with layers) the analysis of the source of relevant FIE is much more complicated and needs to be done in the η, m and w planes observing the GO fields along the interface directions (Daniele and Lombardi 2016; Daniele *et al.* 2017b, 2017c, 2018).

With reference to Figure 5.2.1 and to the parameters of the problem, as described in section 5.8: $\Phi = 0.55$, impedances $\mathbf{z}_a, \mathbf{z}_b$ as given below, $k = 1 - j10^{-3} m^{-1}$, $\beta = \pi/3$, $\varphi_o = \pi/10$, $E_o = 0V/m, Z_o H_o = 1V/m$.

$$\mathbf{z}_{\mathbf{a}} = \begin{bmatrix} 2 & 1+0.5j \\ -0.5 & 10 \end{bmatrix}, \mathbf{z}_{\mathbf{b}} = \begin{bmatrix} 10 & 0 \\ 0 & 0.1 \end{bmatrix} \tag{5.14.1}$$

Matrix (5.14.1) follows the constraint of passivity (Zhu and Landstorfer 1997).

The approximate solutions obtained with the Fredholm factorization method are obtained with discretization parameters A = 10 and h = 0.5.

By analyzing GO waves using geometrical optics or using the spectral domain w, we have seven poles $w_{oi} = (-3.614, -1.886, -1.414, 0.314, 0.786, 2.514, 2.99)$ related to the incident wave and to the single-, double- and triple-reflected waves. The waves' coefficients are properly evaluated using the residue formula and/or GO considerations. Only the triple-reflected waves generate shadow boundaries at $\varphi = \mp 6\Phi - \varphi_o \pm \pi$. Figure 5.14.2 shows the total fields at $k\rho = 10$ for the co-polar and cross-polar components.

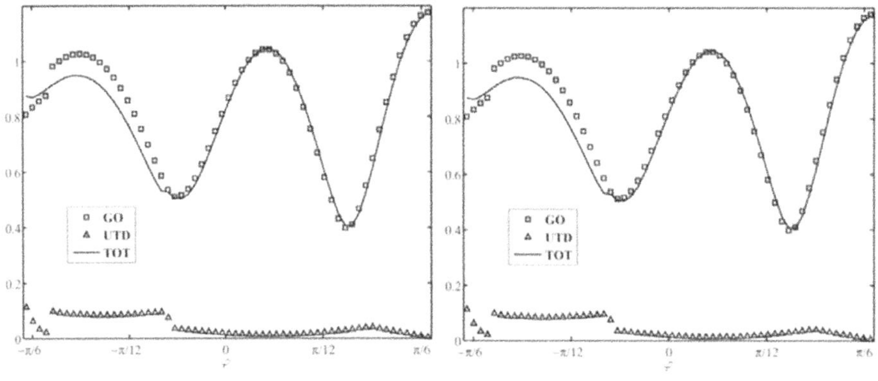

Figure 5.14.2. *Co-polar H_z (left) and cross-polar E_z/Z_o (right) far-field components: total field (solid line), GO field component (squares) and UTD field component (triangles). Note the shadow boundaries at $\varphi = \mp 6\Phi - \varphi_o \pm \pi$ of the tripled reflected waves (© 2012 IEEE[4])*

Another interesting test is the application of our methodology when no diffractive component is excited. This is the case of the right-angled concave wedge with the compatibility conditions on the face impedances reported in Senior and Osipov (2007).

4 Reprinted with permission from G. Lombardi, "Skew incidence on concave wedge with anisotropic surface impedance," *IEEE Antennas Wireless Propag. Lett.*, vol. 11, pp. 1141–1145, Sep. 2012.

By selecting $\Phi = \pi/4$, impedances $\mathbf{z}_a, \mathbf{z}_b$ as given below, $k = 1 - j10^{-3} m^{-1}$, $\beta = \pi/4$, $\varphi_o = \pi/6$, $E_o = 1V/m, H_o = 0A/m$, the results in terms of GTD diffraction coefficient obtained with the Fredholm factorization method (A = 10, h = 0.5), are presented in Figure 5.14.3, which validate the compatibility condition:

$$\mathbf{z_a} = \begin{bmatrix} 1 & 0 \\ 0 & 0.2 \end{bmatrix}, \mathbf{z_b} = \begin{bmatrix} 1.25 & 0 \\ 0 & 1 \end{bmatrix} \qquad (5.14.2)$$

In fact, from Figure 5.14.3, we note GTD diffraction (co-polar and cross-polar) coefficients of approximately −300 dB.

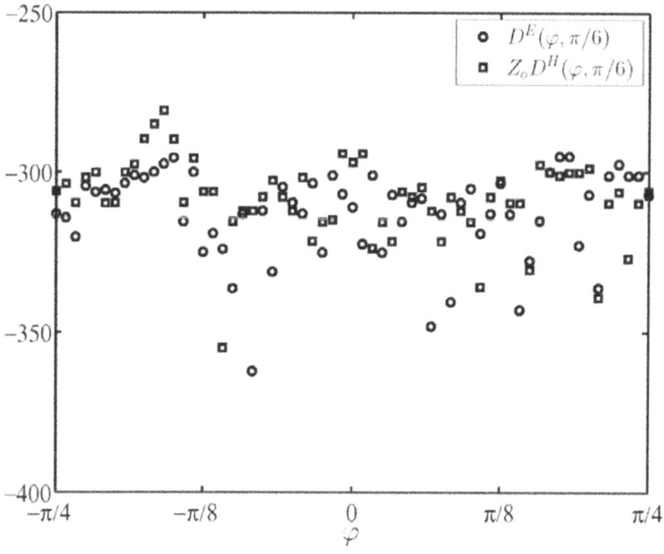

Figure 5.14.3. *Co-polar (E_z) and cross-polar (Z_o H_z) GTD diffraction coefficients in dB for the test case of section 5.14 implementing the compatibility conditions on the face impedances. The result is that the diffractive component is vanishing (© 2012 IEEE[5])*

5 Reprinted with permission from G. Lombardi, "Skew incidence on concave wedge with anisotropic surface impedance," *IEEE Antennas Wireless Propag. Lett.*, vol. 11, pp. 1141–1145, Sep. 2012.

5.15. Discussion

The GWHE formulation together with the direct application of the Fredholm factorization method in $\overline{w}$ constitutes a tool to solve general arbitrary impenetrable wedge problems.

In particular, the method is capable of handling the scattering problem of a plane wave with arbitrary incidence and polarization on an anisotropic non-symmetric wedge with any aperture angle.

In this chapter, we have shown the capability of this method with an in-depth analysis on the analytic properties of functions, as well as the performance of the spectral approximations. Practical problems of scattering by impenetrable wedge are presented for physical and engineering applications. The proposed examples are shown in detail to enable independent implementation by the reader.

Appendix 5.A. Fredholm properties of the integral equation (5.3.1)

In this Appendix, we briefly justify that (5.3.1) is a Fredholm integral equation of the second kind, and thus we obtain good convergence properties of its approximations.

First of all, if we approximate (5.3.1) with an equation where the integration interval is limited (e.g. $-A<\overline{\eta}<A$ with real $\overline{\eta}$), provided A is sufficiently large, the solution of this approximate problem is very close to the exact solution. In fact, taking into account Meixner's condition near the edge of the wedge and Watson's lemma, we have

$$X_+(\eta)=\begin{vmatrix} V_{z+}(\eta,0) \\ V_{\rho+}(\eta,0) \\ Z_o I_{z+}(\eta,0) \\ Z_o I_{\rho+}(\eta,0) \end{vmatrix}=\begin{vmatrix} O(\eta^{-(c+1)}) \\ O(\eta^{-c}) \\ O(\eta^{-(c+1)}) \\ O(\eta^{-c}) \end{vmatrix},\ c>0,\ \text{as}\ \eta\to\infty \tag{5.A.1}$$

It allows us to claim that we are dealing with an unknown that in practice is negligible outside the interval $-A<\overline{\eta}<A$. Due to its smoothness, the kernel of (5.3.1) integrated in the finite range $-A<\overline{\eta}<A$ is compact and it guarantees the convergence of the numerical solution via quadrature (Kantorovich and Krylov 1964); hence, we can correctly label Fredholm the integral equation (5.3.1) when it is defined in the limited interval $-A<\overline{\eta}<A$ for any finite value of A.

A rigorous validation that (5.3.1) is a Fredholm integral equation of the second kind needs to prove the boundedness of the norm of the integral operator in $-\infty < \bar{\eta} < \infty$.

To ascertain this property, we need to introduce a suitable space of functions where we define (5.3.1). We note that a generalized Hilbert space $M_\mu = L_2(\mathbb{R}, \mu(\bar{\eta}))$ (as suggested in Budaev (1995) for operators defined in the w plane) is adequate to show that the kernel of (5.3.1) is quadratically integrable, where the weight $\mu(\bar{\eta})$ introduces a different definition of the scalar product with respect to the classical Hilbert space $L_2(\mathbb{R})$

However, it is remarkable that all the properties of the space $L_2(\mathbb{R})$ also hold in the space M_μ. In particular, the kernel in the integral equation (5.3.1) is compact while selecting, for example, the weight function $\mu(\bar{\eta}) = 1/\left(1+\frac{\bar{\eta}^2}{\tau_o^2}\right)^\alpha$ where α depends on the geometrical and physical parameters of the wedge problem and it is related to the asymptotic behaviors of the kernel and of the integrand (Meixner's conditions).

This proof has already been applied to more complex scattering problems constituted of layers and angular regions with the FIE originating from GWHEs (Daniele *et al.* 2017a).

6

Diffraction by Penetrable Wedges

6.1. Introduction

While solutions in the static case are well known (Scharstein 1993), the only cases that allow exact dynamic solution of penetrable wedges are constituted of the isorefractive wedge (Knockaert *et al.* 1997; Daniele 2003a; Daniele and Uslenghi 2007; Daniele and Zich 2014) and the ideal double-negative (DNG) wedge (Daniele *et al.* 2020). Conversely, the wedge problems constitute a very difficult problem when more than one wave number is present. In fact, in spite of a lot of effort by several authors, the wedge problem with multiple wave numbers is unsolved in closed form.

The literature reports a large number of works about the diffraction of the penetrable wedge. This problem is studied in this chapter in the context of electromagnetics (extendable to other physics) using the Fredholm factorization method applied to GWHEs. An efficient semi-analytical solution is presented in this chapter (Daniele 2009, 2010, 2011a, 2011b; Daniele and Lombardi 2011).

The WH formulation for the penetrable wedge problems presented in this chapter differs from the one used by Croisille and Lebeau in Croisille and Lebeau (1999). These authors represent the unknowns with one-dimensional Laplace transforms and reduce the problem to the solution of spectral integral equations. Conversely, the integral equations of Croisille and Lebeau (1999) derived using the equivalence theorem in the context of elasticity are, and they are not Fredholm integral equations of the second kind (FIE) for the presence of singular kernels.

This book consists of two volumes; any references to Chapters 1, 2 or 3 can be found in the first volume: *Scattering and Diffraction by Wedges 1*.

In Budaev (1995), the author proposed the application of the popular and effective Sommerfeld–Malyuzhinets (SM) method to deal with the planar elastic isotropic wedge scattering problem. The difference equations that arise from this formulation are initially reduced to singular integral equations. A successive regularization method produces the Fredholm equations of the second kind. However, Budaev focused his monograph on the correct mathematical formulations, avoiding engineering aspects such as the evaluation of the diffraction coefficients.

Rawlins (1977) provided a solution of the right dielectric wedge. This solution is based on a perturbation technique applied to an integral equation formulation. His method was generalized in Kim *et al.* (1991a, 1991b). Formulations in the one-dimensional spectral domain based on the Kontorovich–Lebedev (KL) transform have also been proposed in Rawlins (1999) and Salem *et al.* (2006).

In this context, a very important problem is the diffraction by an isotropic elastic wedge. Using the proposed WH technique, this problem is solved when the aperture angle is 360 degrees, i.e. the half-infinite crack (sections 2.11–2.12), and in the relatively simple case of an elastic solid wedge with anti-planar shear motion (section 3.15).

The general scattering problem of an elastic solid wedge presents the same mathematical difficulties as the electromagnetic dielectric wedge (Budaev 1995; Croisille and Lebeau 1999; Gautesen 2001; Gautesen and Fradkin 2010).

As far as concerns the WH solution of the elastic solid wedge scattering problem, it is important to generalize equations (3.3.57)–(3.3.58). This has been accomplished in Daniele (2004) where the GWHEs were obtained in presence of arbitrary linear media. The application of the method to the elastic wedge is currently in progress.

Before considering the details on the dielectric wedge WH solution, we recall the procedure we followed for the Fredholm factorization of GWHEs relevant to an arbitrary impenetrable wedge (Chapters 3 and 5). Referring to Figure 3.3.2, with regions 3 and 4 being impenetrable, the first step is to consider the functional equations only for angular regions 1 and 2 (3.3.57)–(3.3.60). In these equations, the spectra (say $X_+(\eta)$ and $Y_+(-m)$) are plus and minus functions in different complex planes respectively η and m that are related to one another. To make these spectra defined in the same plane with their regularity properties (plus and minus), we introduced (second step) the mapping $\eta(\bar{\eta})$ (3.4.1). Section 3.4 of Chapter 3 shows that $\bar{X}_+(\bar{\eta}) = X_+(\eta(\bar{\eta}))$ and $\bar{X}_-(\bar{\eta}) = Y_+(-m(\bar{\eta}))$ are effectively plus and minus functions in $\bar{\eta}$. For impenetrable wedges assuming $\gamma_a = \gamma_b = \Phi$ (Figure 3.3.2), the mapping (3.4.1) in the two angular regions 1 and 2 is the same. In this way, taking into account the boundary conditions on the faces a and b, equations (3.3.57)–(3.360) become GWHEs that, using (3.4.1), reduce to classical WH equations in the $\bar{\eta}$ plane (see also compact formulations and derivations in Chapter 5). The Fredholm factorization (third step) is fundamental when exact classical factorization is not available. This factorization is based on the systematic elimination of the minus (or alternatively plus) functions present in all the WH equations of the problem by using Cauchy decompositions (section 1.9 of Chapter 1). The elimination of the minus (or alternatively plus) functions produces the FIE of the second kind in the plus (or alternatively minus) functions. In order to see what the novelties of the WH procedure for the penetrable wedge are, let us consider in particular the dielectric wedge in electromagnetic scattering (Figure 6.1.1).

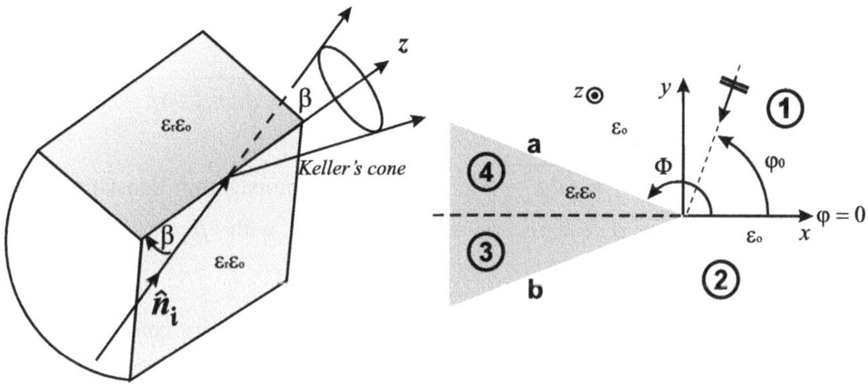

Figure 6.1.1. *Dielectric wedge problem: $2\pi - 2\Phi$ is the aperture angle of the wedge made of dielectric material with relative permittivity ε_r*

Figure 6.1.1 presents the scattering problem where Cartesian coordinates together with cylindrical coordinates are used to describe the problem. We assume the wedge with cylindrical symmetry with respect to the z axis. For simplicity and without loss of generality, we orient the coordinate systems such that the x axis bisects the aperture angle ($2\pi - 2\Phi$) of the wedge.

The dielectric wedge has relative permittivity ε_r and is immersed in free space with permittivity ε_o and permeability μ_o. The problem is characterized by the propagation constant $k_o = \omega\sqrt{\varepsilon_o\mu_o}$ in free space and $k_1 = \sqrt{\varepsilon_r}k_o$ in the dielectric medium and impedances $Z_o = \sqrt{\mu_o / \varepsilon_o}$, $Z_1 = Z_o / \sqrt{\varepsilon_r}$. In the following, we will use the aperture angle $\Phi_1 = \pi - \Phi$.

In the dielectric wedge problem, there are four angular regions to be considered. We need to use all equations (3.3.57)–(3.3.64).

Immediately, we note that although η is the same in the two media (regions 1 and 2 with respect to regions 3 and 4), m is defined in a different way in free space/air (regions 1 and 2) with respect to the dielectric (regions 3 and 4).

In fact, we have respectively in free space and dielectric medium:

$$m = m(\eta) = -\eta \cos \Phi + \xi(\eta) \sin \Phi \tag{6.1.1}$$

$$m_d = m_d(\eta) = -\eta \cos \Phi_1 + \xi_d(\eta) \sin \Phi_1 \tag{6.1.2}$$

where:

$$\Phi_1 = \pi - \Phi, \quad \xi = \xi(\eta) = \sqrt{\tau_o^{\ 2} - \eta^2}, \quad \xi_d = \xi_d(\eta) = \sqrt{\varepsilon_r \tau_o^{\ 2} - \eta^2} \tag{6.1.3}$$

with $\tau_o = \sqrt{k^2 - \alpha_o^2}$, $\tau_d = \sqrt{\varepsilon_r k^2 - \alpha_o^2}$ and k is the propagation constant of free space with $\alpha_o = k \cos \beta$ (β is the skewness angle with $\beta = \pi/2$ at normal incidence).

The different m produces a difficulty in imposing the boundary conditions on the interfaces a and b. We remember that in the natural domain, this is very simple: just the continuity of the tangential component of the electromagnetic field.

For example, on face a, we have $E_z(\rho, \Phi_+) = E_z(\rho, \Phi_-)$. The two quantities $E_z(\rho, \Phi_+), E_z(\rho, \Phi_-)$ in the spectral domain are defined with spectra $V_{z+}(-m(\eta), \Phi_+)$, $V_{z+}(-m_d(\eta), \Phi_-)$, which for the same value of η gives different values.

To overcome this difficulty, in some steps of the WH procedure, we use the Cauchy formula that introduces an integral representation between $V_{z+}(-m(\eta), \Phi_+)$ and $V_{z+}(-m_d(\eta), \Phi_-)$.

To apply the Fredholm factorization method, we must use the mapping (3.4.1) that, for a given angular region, produces plus and minus functions in the same complex plane.

Of course, we have two different mappings for the free space/air (regions 1 and 2) and dielectric medium (regions 3 and 4). In fact, we have:

$$\eta = -\tau_o \cos\left[\frac{\Phi}{\pi}\arccos(-\frac{\bar{\eta}}{\tau_o})\right] \text{ for the free space/air} \tag{6.1.4}$$

$$\eta = \eta_d = -\tau_d \cos\left[\frac{\Phi_1}{\pi}\arccos(-\frac{\bar{\eta}_d}{\tau_d})\right] \text{ for the dielectric medium} \tag{6.1.5}$$

where $\tau_o = \sqrt{k^2 - \alpha_o^2}$, $\tau_d = \sqrt{\varepsilon_r k^2 - \alpha_o^2}$.

In the following, we summarize the procedure to obtain the solution.

Using (6.1.4) and (6.1.5), we rewrite (3.3.57)–(3.3.60) in the $\bar{\eta}$ plane and (3.3.60)–(3.3.64) in the $\bar{\eta}_d$ plane. After algebraic normalizations and manipulations, the equations are written in a form that the Cauchy decompositions immediately eliminate the plus function, yielding integral equations with only minus unknown functions defined in $\bar{\eta}$ and $\bar{\eta}_d$.

Before the application of the boundary conditions on faces a and b, we transform the integral equations respectively in the w plane and $\bar{w} = \frac{\pi}{\Phi} w$ for the η and $\bar{\eta}$ planes and w_d and $\bar{w}_d - \frac{\pi}{\Phi_1} w_d$ for the $\eta_d = \eta$ and $\bar{\eta}_d$ planes.

The substitutions $\bar{w} = -\frac{\pi}{2} + ju$, $\bar{w}_d = -\frac{\pi}{2} + jv$ reduce the integration lines to real axis of u and v.

Finally, we impose the boundary conditions using a suitable mapping and an integral representation derived by the Cauchy formula.

With this procedure, we obtain a system of integral equations on the real axis u that allows the evaluation of the minus spectra on this line.

From Chapter 3, we recall that the minus spectra represent the Laplace transform of the tangential components on the faces and the plus spectra represent the Laplace transforms on the field on the axis x (axial spectra).

The use of axial spectra is more suitable to evaluate the field in the spectral domain for any observation angle φ (3.10.5).

Therefore, since the integral equations are written in minus spectra, once solved, we use the GWHE to obtain the plus spectra.

Analytic continuations of the plus spectra provide the Laplace transform of all the components of the electromagnetic field for any observation angle φ.

It is remarkable that despite the intricate procedure, all the described steps involve elementary mathematical tools.

Furthermore, the kernel of the integral equations written in terms of minus spectra is very simple and suitable for approximate semi-analytical solutions as for the impenetrable wedge problems (Chapter 5).

From a conceptual point of view, there is no difference between the normal incidence and the skew incidence. However, the skew incidence doubles the number of the unknown spectra.

To make the exposition clearer, it is convenient to start to consider the normal incidence case.

In the following, we denote parameters and functions relevant to the dielectric medium with label 1 or d, while parameters of the free space/air with no label or o.

6.2. GWHEs for the dielectric wedge at normal incidence (Ez-polarization)

For the sake of simplicity, in this section, we consider a plane wave at normal incidence, i.e. the wave vector lie in a plane whose perpendicular line is the edge of the wedge. The general case of skew incidence is considered in section 6.3.

Figure 6.1.1 illustrates the problem of the diffraction by a plane wave on a dielectric wedge having relative permittivity ε_r immersed in free space with permittivity ε_o and permeability μ_o.

We consider time harmonic electromagnetic fields with a time dependence specified by the factor $e^{j\omega t}$, which is omitted. The incident field is constituted of the Ez-polarized plane wave having the following longitudinal component:

$$E_z^i = E_o e^{jk_o \rho\cos(\varphi-\varphi_o)} \tag{6.2.1}$$

where φ_o is the azimuthal angle of incidence and $k_o = \omega\sqrt{\mu_o \varepsilon_o}$. The two different angular homogeneous regions of Figure 6.1.1 are defined by $0<|\varphi|<\Phi$

(free space) and $\Phi<|\varphi|<\pi$ (dielectric wedge). For simplicity, in this chapter, we consider Φ obtuse: $\Phi>\pi/2$.

As discussed in the previous chapters, the Wiener-Hopf technique for wedge problems is based on the introduction of the following Laplace transforms in the η plane:

$$V_{z+}(\eta,\varphi)=\int_0^\infty E_z(\rho,\varphi)e^{j\eta\rho}d\rho \tag{6.2.2}$$

$$I_{\rho+}(\eta,\varphi)=\int_0^\infty H_\rho(\rho,\varphi)e^{j\eta\rho}d\rho \tag{6.2.3}$$

At normal incidence $\alpha_o=0$ and for Ez-polarization, the fields H_z, E_ρ and E_φ are null.

Equations (3.3.58), (3.3.60), (3.3.62) and (3.3.64) are automatically satisfied, and equations (3.3.57), (3.3.59), (3.3.61) and (3.3.63) become:

$$\xi\, V_{o+}(\eta)-\omega\mu_o I_{o+}(\eta)=-n\, V_{a+}(-m)-\omega\mu_o I_{a+}(-m) \tag{6.2.4}$$

$$\xi V_{o+}(\eta)+\omega\mu_o I_{o+}(\eta)=-n\, V_{b+}(-m)+\omega\mu_o I_{b+}(-m) \tag{6.2.5}$$

$$\xi_1\, V_{\pi+}(\eta)+\omega\mu_o I_{\pi+}(\eta)=-n_1\, V_{a+}(-m_1)+\omega\mu_o I_{a+}(-m_1) \tag{6.2.6}$$

$$\xi_1 V_{\pi+}(\eta)-\omega\mu_o I_{\pi+}(\eta)=-n_1\, V_{b+}(-m_1)-\omega\mu_o I_{b+}(-m_1) \tag{6.2.7}$$

where:

$$V_{a,b+}(-m)=V_{z+}(-m,\pm\Phi),\quad I_{a,b+}(-m)=I_{\rho+}(-m,\pm\Phi),$$

$$V_{o+}(\eta)=V_{z+}(\eta,0),\quad I_{o+}(\eta)=I_{\rho+}(\eta,0),$$

$$V_{\pi+}(\eta)=V_{z+}(\eta,\pi),\quad I_{\pi+}(\eta)=I_{\rho+}(\eta,\pi)$$

with $\Phi_1=\pi-\Phi$, $\xi=\xi(\eta)=\sqrt{k_o^2-\eta^2}$, $\xi_1=\xi_1(\eta)=\sqrt{\varepsilon_r k_o^2-\eta^2}$,

$$m=m(\eta)=-\eta\cos\Phi+\xi\sin\Phi,\quad m_1=m_1(\eta)=-\eta\cos\Phi_1+\xi_1\sin\Phi_1$$

$$n=n(\eta)=-\xi\cos\Phi-\eta\sin\Phi,\quad n_1=n_1(\eta)=-\xi_1\cos\Phi_1-\eta\sin\Phi_1$$

The multivalued functions ξ and ξ_1 define proper sheets with $\xi(0)=k_o$, $\xi_1(0)=k_1=\sqrt{\varepsilon_r}k_o$ and we assume as standard branch lines the ones with null imaginary parts of ξ and ξ_1.

In (6.2.4)–(6.2.7), the unknowns are plus functions in the η plane or minus functions in m, m_1 respectively (6.2.4)–(6.2.5) and (6.2.6)–(6.2.7).

6.3. Reduction of the GWHEs for the dielectric wedge at Ez-polarization to Fredholm integral equations

The GWHEs (6.2.4)–(6.2.7) are not written in normal form. To obtain normal forms, we operate the factorization of the scalars ξ, ξ_1, n, n_1, i.e. $\xi=\xi_-\xi_+$, $\xi_1=\xi_{1-}\xi_{1+}$, $n=n_-n_+$, $n_1=n_{1-}n_{1+}$, where the subscript + means plus function in the η plane and the subscript – means minus function in the m, m_1 plane. The explicit expressions of the factorized functions can be obtained as described in Appendix 4.A, i.e. by exploiting the properties of plus and minus functions in the angular complex plane (see section 3.4 after (3.4.9)).

In the $\bar{\eta}$ and $\bar{\eta}_d$ planes defined by mappings (6.1.4)–(6.1.5), we have

$$n_+=\sqrt{\frac{k_o-\bar{\eta}}{2}},\ n_{1+}=\sqrt{\frac{k_1-\bar{\eta}_d}{2}},\ \xi_-=-\sqrt{\frac{k_o+\bar{\eta}}{2}},\ \xi_{1-}=-\sqrt{\frac{k_1+\bar{\eta}_d}{2}} \tag{6.3.1}$$

In the following, we use the dot notation, $\dot{F}$, for quantities defined on the interface between the media with spectral variable m_1 of the dielectric medium. The same quantities are represented without a dot, F, if defined in the spectral variable m of the free space.

Simple algebraic manipulations on (6.2.4)–(6.2.7) yield the following normalized GWHEs of the problem:

$$Y_{1+}(\eta)=X_{1+}(-m)-\frac{\xi_-}{n_+}X_{2+}(-m) \tag{6.3.2}$$

$$Y_{2+}(\eta)=\dot{X}_{1+}(-m_1)+\frac{\xi_{1-}}{n_{1+}}\dot{X}_{2+}(-m_1) \tag{6.3.3}$$

$$Y_{3+}(\eta) = X_{3+}(-m) - \frac{\xi_-}{n_+} X_{4+}(-m) \tag{6.3.4}$$

$$Y_{4+}(\eta) = \dot{X}_{3+}(-m_1) + \frac{\xi_{1-}}{n_{1+}} \dot{X}_{4+}(-m_1) \tag{6.3.5}$$

where:

$$Y_{1+}(\eta) = 2\frac{\xi_+}{n_{o+}} V_{o+}(\eta),\ Y_{2+}(\eta) = 2\frac{\xi_{1+}}{n_{1+}} V_{\pi z+}(\eta),$$

$$X_{1+}(-m) = -\frac{n_{o-}}{\xi_-}\left[V_{a+}(-m\ ) + V_{b+}(-m\ )\right],$$

$$\dot{X}_{1+}(-m_1) = -\frac{n_{1-}}{\xi_{1-}}\left[V_{a+}(-m_1) + V_{b+}(-m_1)\right],$$

$$X_{2+}(-m) = \frac{\omega\mu_o}{\xi_-^2}\left[I_{a+}(-m\ ) - I_{b+}(-m\ )\right],$$

$$\dot{X}_{2+}(-m_1) = \frac{\omega\mu_o}{\xi_{1-}^2}\left[I_{a+}(-m_1) - I_{b+}(-m_1)\right],$$

$$Y_{3+}(\eta) = \frac{2\omega\mu_o}{n_{o+}\sqrt{k_o}} I_{o+}(\eta),\ Y_{4+}(\eta) = -\frac{2\omega\mu_o}{n_{1+}\sqrt{k_1}} I_{\pi+}(\eta)$$

$$X_{3+}(-m) = \frac{n_{o-}}{\sqrt{k_o}}\left[V_{a+}(-m\ ) - V_{b+}(-m\ )\right],$$

$$\dot{X}_{3+}(-m_1) = \frac{n_{1-}}{\sqrt{k_1}}\left[V_{a+}(-m_1) - V_{b+}(-m_1)\right]$$

$$X_{4+}(-m) = -\frac{\omega\mu_o}{\sqrt{k_o}\,\xi_-}\left[I_{a+}(-m\ ) + I_{b+}(-m\ )\right],$$

$$\dot{X}_{4+}(-m_1) = -\frac{\omega\mu_o}{\sqrt{k_1}\,\xi_{1-}}\left[I_{a+}(-m_1) + I_{b+}(-m_1)\right]$$

Now the equations are normal because the first members are just plus functions.

Consequently, the Fredholm factorization in $\bar{\eta}$ and $\bar{\eta}_d$ planes (6.1.4)–(6.1.5) eliminates these unknowns yielding a system of FIE having as unknowns only minus functions (see Appendix 6.A).

We prefer to include the details on derivation of FIEs in Appendix 6.A to focus the attention of the reader on the practical description of the method to solve the scattering problem under consideration.

Note that equations (6.3.2) and (6.3.3) are not coupled with equations (6.3.4) and (6.3.5) that have the same forms.

Let us only focus on systems (6.3.2) and (6.3.3) (rather than (6.3.4) and (6.3.5)).

The procedure consists of the elimination of plus functions $\bar{Y}_{1+}(\bar{\eta}) = Y_{1+}(\eta)$ and $\bar{Y}_{2+}(\bar{\eta}_d) = Y_{2+}(\eta)$ by the Cauchy decomposition, which yields two integral equations involving the minus functions $\bar{X}_{1+}(-m(\bar{\eta})), \dot{\bar{X}}_{1+}(-m_d(\bar{\eta}_d))$ and $\bar{X}_{2+}(-m(\bar{\eta}))$, $\dot{\bar{X}}_{2+}(-m_d(\bar{\eta}_d))$.

As considered in Chapters 1 and 5, it is convenient to introduce the complex plane $\bar{w}$ and $\bar{w}_d$ defined by

$$\bar{\eta} = -k\cos\bar{w}, \quad \bar{\eta}_d = -k_d\cos\bar{w}_d \tag{6.3.6}$$

In particular, we apply the warping of the integration lines on the straight lines $u \in \mathbb{R}$ and $v \in \mathbb{R}$ defined by:

$$\bar{w} = -\frac{\pi}{2} + ju\,, \quad \bar{w}_d = -\frac{\pi}{2} + jv \tag{6.3.7}$$

Finally, we obtain from (6.3.2) and (6.3.3) via Appendix 6.A

$$P_1(u)-\tan\frac{-\frac{\pi}{2}+ju}{2}P_2(u)-\frac{1}{\pi j}\frac{e^u}{\left(e^u-j\right)}\int_{-\infty}^{\infty}\frac{e^{u'}+j}{e^{u+u'}+1}P_2(u')du'=-\frac{T_o(w(u))}{(j\sinh u-\cos\frac{\pi}{\Phi}\varphi_o)},u\in\mathbb{R}$$

$$e_{Q_1}(v)=Q_1(v)+\tan\frac{-\frac{\pi}{2}+jv}{2}Q_2(v)+\frac{1}{\pi j}\frac{e^v}{\left(e^v-j\right)}\int_{-\infty}^{\infty}\frac{e^{v'}+j}{e^{v+v'}+1}Q_2(v')dv'=0,v\in\mathbb{R}$$

(6.3.8)

where for $i=1,2,$ we have

$$P_i(u)=k_oX_{i+}(-k_o\cos\left[\frac{\Phi}{\pi}\left(-\frac{\pi}{2}+ju\right)+\Phi\right]),$$
$$Q_i(v)=k_o\dot{X}_{i+}(-k_1\cos\left[\frac{\Phi_1}{\pi}\left(-\frac{\pi}{2}+jv\right)+\Phi_1\right]) \quad (6.3.9)$$

with propagation constant $k_o=\omega\sqrt{\varepsilon_o\mu_o}$ in free space and $k_1=\sqrt{\varepsilon_r}k_o$ in the dielectric medium, impedances $Z_o=\sqrt{\mu_o/\varepsilon_o}$ and $Z_1=Z_o/\sqrt{\varepsilon_r}$, and aperture angle $\Phi_1=\pi-\Phi$.

Repeating the same procedure for (6.3.4) and (6.3.5), where the plus functions are $\overline{Y}_{3+}(\overline{\eta})=Y_{3+}(\eta)$ and $\overline{Y}_{4+}(\overline{\eta}_d)=Y_{4+}(\eta)$, and the minus functions are $\overline{X}_{3+}(-m(\overline{\eta})),\dot{\overline{X}}_{4+}(-m_d(\overline{\eta}_d))$ and $\overline{X}_{3+}(-m(\overline{\eta})),\dot{\overline{X}}_{4+}(-m_d(\overline{\eta}_d))$, we obtain

$$P_3(u)-\tan\frac{-\frac{\pi}{2}+ju}{2}P_4(u)-\frac{1}{\pi j}\frac{e^u}{\left(e^u-j\right)}\int_{-\infty}^{\infty}\frac{e^{u'}+j}{e^{u+u'}+1}P_4(u')du'=$$
$$-\frac{U_o(w(u))}{(j\sinh u-\cos\frac{\pi}{\Phi}\varphi_o)},u\in\mathbb{R} \quad (6.3.10)$$

$$e_{Q_3}(v)=Q_3(v)+\tan\frac{-\frac{\pi}{2}+jv}{2}Q_4(v)+\frac{1}{\pi j}\frac{e^v}{\left(e^v-j\right)}\int_{-\infty}^{\infty}\frac{e^{v'}+j}{e^{v+v'}+1}Q_4(v')dv'=0,v\in\mathbb{R}$$

where we use definitions (6.3.9) for $i=3,4$.

The source term $T_o(w)$ ($U_o(w)$) derives from the non-standard contributions of the GO field present in the problem (1.9.7). A detailed discussion of these contributions is presented in Appendix 6.B (generalizations are reported in Daniele and Lombardi (2011)) with slightly different notations. Moreover, in this case, we prefer to include the details on derivation of the source of the FIEs in an Appendix to focus the attention of the reader on the description of the method.

It yields:

$$T_o(w) = \begin{cases} -j4E_o\dfrac{\pi}{\Phi}, & 0 \le \varphi_o < \Phi/2 \\ \bar{R}_{1o} - \tan\left(\dfrac{\pi w}{2\Phi}\right)\bar{R}_{2o}, & \Phi/2 < \varphi_o < \Phi \end{cases} \tag{6.3.11}$$

$$U_o(w) = \begin{cases} -j4E_o\dfrac{\pi}{\Phi}\sin(\dfrac{\pi}{2\Phi}\varphi_o), & 0 \le \varphi_o < \Phi/2 \\ \bar{R}_{3o} - \tan\left(\dfrac{\pi w}{2\Phi}\right)\bar{R}_{4o}, & \Phi/2 < \varphi_o < \Phi \end{cases} \tag{6.3.12}$$

with

$$\bar{R}_{1o} = -2j\frac{\pi}{\Phi}(1+\Gamma_e)E_o,\ \bar{R}_{2o} = -2j\frac{\pi}{\Phi}(1-\Gamma_e)\cot(\frac{\pi\,\varphi_o}{2\Phi})E_o$$

$$\bar{R}_{3o} = -2j\frac{\pi}{\Phi}(1+\Gamma_e)\sin(\frac{\pi\,\varphi_o}{2\Phi})E_o,\ \bar{R}_{4o} = -2j\frac{\pi}{\Phi}(1-\Gamma_e)\cos(\frac{\pi\,\varphi_o}{2\Phi})E_o$$

and where Γ_e is the reflection coefficients of face a, which is given by

$$\Gamma_e = \frac{Z_1\sin(\Phi-\varphi_o) - Z_o\sin(\Phi-\varphi_{to})}{Z_1\sin(\Phi-\varphi_o) + Z_o\sin(\Phi-\varphi_{to})} \tag{6.3.13}$$

with $\varphi_{to} = \Phi + g(\Phi-\varphi_o)$, and

$$g(w) = -\arccos\left(\frac{\cos w}{\sqrt{\varepsilon_r}}\right) \tag{6.3.14}$$

Systems (6.3.8) and (6.3.10) introduce a number of unknowns which is double the number of available equations at this moment.

In fact, we still need to impose the boundary conditions on the faces a and b. For this item, we focus on system (6.3.8). We note that an analytic continuation of the integral in the second equation of (6.3.8) makes $e_{Q_1}(v)=0$ valid for any value of v, as well as outside the real axis, until the singularity lines of the kernel $M(v,v')=\frac{1}{\pi j}\frac{e^{v}}{\left(e^{v}-j\right)}\frac{e^{v'}+j}{e^{v+v'}+1}$ are encountered (Gakhov 1966). The concept of kernel singularity lines is already used in sections 5.5–5.6 for impenetrable wedge problems and introduced in Daniele (2017a).

Hence, it is convenient to assume values of $v=v(u)$ such that $m=m_1$ thus

$$k_1\cos\left[\frac{\Phi_1}{\pi}\left(-\frac{\pi}{2}+jv\right)+\Phi_1\right]=k_o\cos\left[\frac{\Phi}{\pi}\left(-\frac{\pi}{2}+ju\right)+\Phi\right] \tag{6.3.15}$$

Equation (6.3.15) defines a line (6.3.16) of the plane w that for $u\in R$ that does not cross any singularity line of the kernel $M(v,v')$:

$$v=v(u)=\frac{j\pi}{2\Phi_1}\left(\Phi_1-2\arccos\frac{\cos\frac{(\pi+2ju)\Phi}{2\pi}}{\sqrt{\varepsilon_r}}\right) \tag{6.3.16}$$

Applying an analytic continuation, the second equation of (6.3.8) can be rewritten: ($u\in R$)

$$Q_1(v(u))+\tan\frac{-\frac{\pi}{2}+jv}{2}Q_2(v(u))+\frac{1}{\pi j}\frac{e^{v}}{\left(e^{v}-j\right)}\int_{-\infty}^{\infty}\frac{e^{v'}+j}{e^{v+v'}+1}Q_2(v')dv'=0, u\in\mathbb{R} \tag{6.3.17}$$

However, even though we substitute the second equation of (6.3.8) with (6.3.17), we still have two equations with four unknowns.

Now it is worth applying the boundary conditions on the faces a and b of the wedge.

In fact, thanks to (6.3.15), using $v(u)$ given by (6.3.16), the boundary conditions can be enforced as

$$V_{a,b+}(-m\ ) = \dot{V}_{a,b+}(-m_1),\ I_{a,b+}(-m\ ) = \dot{I}_{a,b+}(-m_1) \tag{6.3.18}$$

Taking into account the definitions of $X_{i+}, \dot{X}_{i+}$ (i = 1,2), it yields two equations that link the unknowns $Q_{1,2}(v(u))$ with $P_{1,2}(u)$ for ($u \in R$):

$$Q_1(v(u)) = \frac{\cosh(u)\sin[-\dfrac{(-jv(u)+\frac{\pi}{2})\Phi_1}{\pi}+\Phi_1]}{\cosh(v(u))\sin[-\dfrac{(-ju+\frac{\pi}{2})\Phi}{\pi}+\Phi]}P_1(u) \tag{6.3.19}$$

$$Q_2(v(u)) = \frac{Z_1 \sin^2[\frac{1}{2}(-\frac{\pi}{2}+ju)]}{Z_o \sin^2[\frac{1}{2}(-\frac{\pi}{2}+jv(u))]}P_2(u) \tag{6.3.20}$$

Now (6.3.17) can be rewritten for $u \in R$

$$\frac{\cosh(u)\sin[-\frac{(-jv(u)+\frac{\pi}{2})\Phi_1}{\pi}+\Phi_1]}{\cosh(v(u))\sin[-\frac{(-ju+\frac{\pi}{2})\Phi}{\pi}+\Phi]}P_1(u)+\tan\frac{-\frac{\pi}{2}+jv}{2}\frac{Z_1\sin^2[\frac{1}{2}(-\frac{\pi}{2}+ju)]}{Z_o\sin^2[\frac{1}{2}(-\frac{\pi}{2}+jv(u))]}P_2(u)+\frac{1}{\pi j}\frac{e^v}{\left(e^v-j\right)}\int_{-\infty}^{\infty}\frac{e^{v'}+j}{e^{v+v'}+1}Q_2(v')dv'=0 \tag{6.3.21}$$

Note that in (6.3.21), there is still the presence of $Q_2(v')$ inside the integral for real v', while for $v'(u)$, it is not real for $u \in R$.

For this task, we use the following representation that derives from the Cauchy formula (Appendix of Daniele (2010)):

$$Q_2(v') = \int_{-\infty}^{\infty} T_2(v',u')P_2(u')du' \tag{6.3.22}$$

where:

$$T_2(v,u) = -\frac{\Phi}{2\pi^2}\frac{Z_1 \sin^2[\frac{1}{2}(-\frac{\pi}{2}+ju)]}{Z_o \sin^2[\frac{1}{2}(-\frac{\pi}{2}+jv)]}\frac{\sin[\frac{\Phi}{\pi}(ju+\frac{\pi}{2})+\Phi]}{\cos[\frac{\Phi}{\pi}(ju+\frac{\pi}{2})+\Phi]-\sqrt{\varepsilon_r}\cos[\frac{\Phi_1}{\pi}(jv+\frac{\pi}{2})+\Phi_1]} \tag{6.3.23}$$

From (6.3.8), rewriting (6.3.21) by taking into account (6.3.22) and interchanging the order of integration, we obtain the system (6.3.24) for $u \in R$ (see full page system on page 151).

In system (6.3.24)

$$\begin{aligned} M(u,u') &= \frac{1}{\pi j}\frac{e^u}{(e^u - j)}\frac{e^{u'}+j}{e^{u+u'}+1}, \\ M_{2e}(u,u') &= \int_{-\infty}^{\infty}\frac{\cosh(v(u))}{\cosh(u)}dv' M(v(u),v')T_2(v',u') \end{aligned} \tag{6.3.25}$$

Finally, now we are facing two coupled classical FIEs that are defined for $u \in R$.

The same procedure is applied to system (6.3.10), and it yields equations (6.3.26) for the unknowns $P_3(u)$ and $P_4(u)$ for $u \in R$ (see full page equation on page 151).

$$P_1(u)-\tan\frac{-\frac{\pi}{2}+ju}{2}P_2(u)-\int_{-\infty}^{\infty}M(u,u')P_2(u')du'=-\frac{T_o(w(u))}{(j\sinh u-\cos\frac{\pi}{\Phi}\varphi_o)},$$

$$\frac{\sin[\frac{jv(u)\Phi_1}{\pi}+\frac{\Phi_1}{2}]P_1(u)}{\sin[\frac{ju\Phi}{\pi}+\frac{\Phi}{2}]}+\tan\frac{-\frac{\pi}{2}+jv(u)}{2}\frac{\cosh(v(u))}{\cosh(u)}\frac{Z_1\sin^2[\frac{1}{2}(-\frac{\pi}{2}+ju)]P_2(u)}{Z_o\sin^2[\frac{1}{2}(-\frac{\pi}{2}+jv(u))]}+\int_{-\infty}^{\infty}M_{2e}(u,u')P_2(u')du'=0$$

(6.3.24)

$$P_3(u)-\tan\frac{-\frac{\pi}{2}+ju}{2}P_4(u)-\int_{-\infty}^{\infty}M(u,u')P_4(u')du'=-\frac{U_o(w(u))}{(j\sinh u-\cos\frac{\pi}{\Phi}\varphi_o)},$$

$$\frac{\sin[\frac{jv(u)\Phi_1}{\pi}+\frac{\Phi_1}{2}]P_3(u)}{\sin[\frac{ju\Phi}{\pi}+\frac{\Phi}{2}]}+\tan\frac{-\frac{\pi}{2}+jv(u)}{2}\frac{\cos[\frac{1}{2}(-\frac{\pi}{2}+jv(u))]}{\cos[\frac{1}{2}(-\frac{\pi}{2}+ju)]}\frac{Z_1\sin[\frac{1}{2}(-\frac{\pi}{2}+ju)]P_4(u)}{Z_o\sin[\frac{1}{2}(-\frac{\pi}{2}+jv(u))]}+\int_{-\infty}^{\infty}M_{4e}(u,u')P_4(u')du'=0$$

(6.3.26)

where:

$$Q_4(v') = \int_{-\infty}^{\infty} T_4(v',u')P_4(u')du' \tag{6.3.27}$$

$$T_4(v,u) = -\frac{\Phi}{2\pi^2} \frac{Z_1 \sin[\frac{1}{2}(-\frac{\pi}{2}+ju)]}{Z_o \sin[\frac{1}{2}(-\frac{\pi}{2}+jv)]} \frac{\sin[\frac{\Phi}{\pi}(ju+\frac{\pi}{2})+\Phi]}{\cos[\frac{\Phi}{\pi}(ju+\frac{\pi}{2})+\Phi]-\sqrt{\varepsilon_r}\cos[\frac{\Phi_1}{\pi}(jv+\frac{\pi}{2})+\Phi_1]}$$

$$M_{4e}(u,u') = \int_{-\infty}^{\infty} \frac{\cos[\frac{1}{2}(-\frac{\pi}{2}+jv(u))]}{\cos[\frac{1}{2}(-\frac{\pi}{2}+ju)]} M(v(u),v')T_4(v',u')dv'$$

(6.3.28)

We recall the definitions of $P_i(u)$ and $Q_i(v)$ provided in (6.3.9).

According to the Fredholm factorization method, the kernels $M(u,u')$, $M_{2e}(u,u')$ and $M_{4e}(u,u')$ in the two final systems (6.3.24) and (6.3.26) are compact.

A direct proof of it is given in Appendix C of Daniele (2011b). An alternative, more restrictive strategy can be developed based on the proof of Appendix 5.A done for the impenetrable wedge kernel. The system (6.3.24), (6.3.26) shows the integral kernels $M(u,u'), M_{2e}(u,u'), M_{4e}(u,u')$. Studying the asymptotic behaviors of the kernels and of the related integrands $P_i(u)$ (Meixner's conditions for $u \to \infty$), we can select a suitable generalized Hilbert space with weight function that allows to show that the kernels are quadratically integrable. With this proof, (6.3.24), (6.3.26) effectively constitute a system of FIEs.

Once we have obtained the auxiliary functions $P_i(u)$, $i=1,2,3,4$ by solving systems (6.3.24) and (6.3.26), we obtain the following representations of the axial (plus) spectra unknowns:

$$\hat{V}_+(w) = V_{z+}(-k_o \cos w, 0) = -\frac{1}{k_o}\frac{\sin(\frac{\pi}{\Phi}w)}{4\sin w}\left\{\int_{-\infty}^{\infty} M\left[-\frac{j\pi}{\Phi}(w+\frac{\Phi}{2}), u'\right]P_2(u')du' - \frac{T_o(w)}{(\cos\frac{\pi}{\Phi}w - \cos\frac{\pi}{\Phi}\varphi_o)}\right\} \tag{6.3.29}$$

$$\hat{I}_+(w) = I_{\rho+}(-k_o \cos w, 0) = \frac{1}{k_o}\frac{\cos(\frac{\pi}{2\Phi}w)}{2Z_o}\left\{\int_{-\infty}^{\infty} M\left[-\frac{j\pi}{\Phi}(w+\frac{\Phi}{2}), u'\right]P_4(u')du' - \frac{U_o(w)}{(\cos\frac{\pi}{\Phi}w - \cos\frac{\pi}{\Phi}\varphi_o)}\right\} \tag{6.3.30}$$

$$\hat{V}_{\pi+}(w_1) = V_{z+}(-k_1 \cos w_1, \pi) = \frac{1}{k_o}\frac{\sin(\frac{\pi}{\Phi_1}w_1)}{4\sin w_1}\int_{-\infty}^{\infty} M\left[-\frac{j\pi}{\Phi_1}(w_1+\frac{\Phi_1}{2}), v'\right]Q_2(v')dv' \tag{6.3.31}$$

$$\hat{I}_{\pi+}(w_1) = I_{\rho+}(-k_1 \cos w_1, \pi) = \frac{1}{k_o}\frac{\cos(\frac{\pi}{2\Phi_1}w_1)}{2Z_1}\int_{-\infty}^{\infty} M\left[-\frac{j\pi}{\Phi_1}(w_1+\frac{\Phi_1}{2}), v'\right]Q_4(v')dv' \tag{6.3.32}$$

where:

$$Q_2(v') = \int_{-\infty}^{\infty} T_2(v',u')P_2(u')du' \qquad (6.3.33)$$

$$Q_4(v') = \int_{-\infty}^{\infty} T_4(v',u')P_4(u')du' \qquad (6.3.34)$$

An approximate solution method is provided in section 6.5.

6.4. Analytic continuation for the solution of the dielectric wedge at Ez-polarization

The representations of $\hat{V}_+(w)$, $\hat{I}_+(w)$ (6.3.29)–(6.3.30) and $\hat{V}_{\pi+}(w_1)$, $\hat{I}_{\pi+}(w_1)$ (6.3.31)–(6.3.32) are valid in strips that comprise the vertical lines $w = \text{Re}[-\Phi/2]$ and $w_1 = \text{Re}[-\Phi_1/2]$ respectively, where the FIEs are solved.

To estimate the width of these strips, it is important to note that the obtained representations of the unknowns in w and w_1 are sectional analytic functions (Gakhov 1966) as they are integral representations.

While moving the observation points w and w_1 out of the vertical lines of integration, we perform an analytic continuation of the representations until a kernel singularity line of the kernel M is crossed (Gakhov 1966).

While crossing the kernel singularity lines, the integrals in the representations become discontinuous in w (w_1).

To compensate the discontinuity, we need to add to the representations (6.3.29)–(6.3.32) suitable sectional functions. With this procedure, we can perform analytic continuations in the whole complex plane w (w_1), as expected; see section 1.12.

Unfortunately, the knowledge of the jumps on the integrals constitutes a difficult problem.

We can overcome these difficulties by using an analytic continuation based on recursive equations obtained by the WH equations of the problem.

To ascertain the domains where (6.3.29)–(6.3.32) are valid, the study of the occurrence of the kernel singularity lines is necessary. Analytic continuation is accomplished in Daniele (2010, 2011a, 2011b) and Daniele and Lombardi (2011). As expected, these equations are valid representations and constitute an analytic element of the axial (plus) spectra in the strips $-\Phi < \mathrm{Re}[w] < 0$ for axial spectral unknowns in free space and $-\Phi_1 < \mathrm{Re}[w_1] < 0$ for axial spectral unknowns in the dielectric.

Taking into account the evenness of the plus spectra, we can immediately extend the validity of (6.3.29)–(6.3.32) in the strips $0 < \mathrm{Re}[w] < \Phi$ and $0 < \mathrm{Re}[w_1] < \Phi_1$ respectively.

To obtain the recursive equations that provide valid representations outside the strips $-\Phi < \mathrm{Re}[w] < 0$ and $-\Phi_1 < \mathrm{Re}[w_1] < 0$ as done in the previous chapters and in particular in Chapter 5, we start from the GWHEs (6.2.4)–(6.2.7) rewritten in the w and w_1 planes. We obtain:

$$\begin{aligned}-2k_o \sin w\, \hat{V}_+(w) = -\omega\mu\left[\hat{I}_{a+}(w+\Phi) - \hat{I}_{b+}(w+\Phi)\right] - \\ k_o \sin(w+\Phi)\left[\hat{V}_{a+}(w+\Phi) + \hat{V}_{b+}(w+\Phi)\right]\end{aligned} \qquad (6.4.1)$$

$$\begin{aligned}2\omega\mu\, \hat{I}_+(w) = \omega\mu\left[\hat{I}_{a+}(w+\Phi) + \hat{I}_{b+}(w+\Phi)\right] + \\ k_o \sin(w+\Phi)\left[\hat{V}_{a+}(w+\Phi) - \hat{V}_{b+}(w+\Phi)\right]\end{aligned} \qquad (6.4.2)$$

$$\begin{aligned}-2k_1 \sin w_1\, \hat{V}_{\pi+}(w_1) = \omega\mu\left[\hat{I}_{a+}(w_1+\Phi_1) - \hat{I}_{b+}(w_1+\Phi_1)\right] - \\ k_1 \sin(w_1+\Phi_1)\left[\hat{V}_{a+}(w_1+\Phi_1) + \hat{V}_{b+}(w_1+\Phi_1)\right]\end{aligned} \qquad (6.4.3)$$

$$\begin{aligned}2\omega\mu\, \hat{I}_{\pi+}(w_1) = \omega\mu\left[\hat{I}_{a+}(w_1+\Phi_1) + \hat{I}_{b+}(w_1+\Phi_1)\right] + \\ k_1 \sin(w_1+\Phi_1)\left[\hat{V}_{a+}(w_1+\Phi_1) - \hat{V}_{b+}(w_1+\Phi_1)\right]\end{aligned} \qquad (6.4.4)$$

where we recall that the face spectra are defined by:

$$\hat{V}_{a,b+}(w) = V_{z+}(-k_o \cos(w), \pm\Phi) \quad \hat{I}_{a,b+}(w) = I_{\rho+}(-k_o \cos(w), \pm\Phi)$$
$$\widehat{V}_{a,b+}(w_1) = V_{z+}(-k_1 \cos(w_1), \pm\Phi), \quad \widehat{I}_{a,b+}(w_1) = I_{\rho+}(-k_1 \cos(w_1), \pm\Phi) \tag{6.4.5}$$

With respect to the recursive equations presented in Chapter 5 for an impenetrable wedge, here we have the contemporary presence of two mapping $w(\eta) = w$ and $w_1(\eta) = w_1$ that are connected through Snell's law in the spectral domain

$$\eta = k_1 \cos(w_1) = k_o \cos(w) \tag{6.4.6}$$

or $w_1 = g(w), \quad w = g_1(w_1)$ with

$$g(w) = -\arccos\left(\frac{\cos w}{\sqrt{\varepsilon_r}}\right), \quad g_1(w_1) = -\arccos\left(\sqrt{\varepsilon_r} \cos w_1\right) \tag{6.4.7}$$

Before proceeding to obtain the analytic continuation of the starting spectra defined in the strips $-\Phi < \text{Re}[w] < \Phi$ (or $-\Phi_1 < \text{Re}[w_1] < \Phi_1$), it is interesting to consider some remarks on $g(w)$ and $g_1(w_1)$. Henceforth, the branches of these multivalued functions are those that yield

$$\sin(g(w)) = -\sqrt{1 - \frac{\cos^2 w}{\varepsilon_r}}, \quad \sin(g_1(w_1)) = -\sqrt{1 - \varepsilon_r \cos^2 w_1} \tag{6.4.8}$$

Besides, in order to avoid the presence of singularities on the real axis w (w_1), we assume ε_r with a negative (vanishing) imaginary part. The branch points of the functions $g(w)$ and $g_1(w_1)$ are defined by taking into account where

$$\frac{dg(w)}{dw} = -\frac{\sin w}{\sqrt{\varepsilon_r - \cos^2 w}}, \quad \frac{dg_1(w_1)}{dw_1} = -\frac{\sqrt{\varepsilon_r} \sin w_1}{\sqrt{1 - \varepsilon_r \cos^2 w_1}} \tag{6.4.9}$$

is null or infinity (practical rule used in section 3.4 to search possible branches and singularities).

These equations yield infinite branch points given by:

$$\pm w_b + n\pi,\ \pm w_{1b} + n\pi \quad \text{with } n = 0, \pm 1, \pm 2, ... \tag{6.4.10}$$

and where:

$$w_b = \arccos\sqrt{\varepsilon_r} = j\text{arccosh}\sqrt{\varepsilon_r},\ \ w_{1b} = -\arccos\frac{1}{\sqrt{\varepsilon_r}} \tag{6.4.11}$$

To simplify the analysis, we select as branches of functions $g(w)$ and $g_1(w_1)$ the ones starting from its branch point and ending to infinity without cutting the real axis. For instance, the branch line relevant to w_b is the vertical line ($w_b, w_b + j\infty$). Other properties of $g(w)$ and $g_1(w_1)$ are described in Appendix II of Daniele and Lombardi (2011).

Snell's law (6.4.6) yields:

$$\hat{V}_{a,b+}(w_1) = \hat{V}_{a,b+}(w),\ \ \hat{I}_{a,b+}(w_1) = \hat{I}_{a,b+}(w) \tag{6.4.12}$$

It follows that if we substitute w with $\pm w - \Phi$ into equations (6.4.1) and (6.4.2) and w_1 with $\pm w_1 - \Phi_1$ into equations (6.4.3) and (6.4.4), we obtain eight independent equations that contain the face spectra $\hat{V}_{a,b+}(\pm w)$ and $\hat{I}_{a,b+}(\pm w)$.

The other unknowns are the axial spectra. Taking into account that all the plus functions are always even functions, we have four additional equations $\hat{V}_{a,b+}(w) = \hat{V}_{a,b+}(-w)$ and $\hat{I}_{a,b+}(w) = \hat{I}_{a,b+}(-w)$. By eliminating the eight unknowns $\hat{V}_{a,b+}(\pm w)$ and $\hat{I}_{a,b+}(\pm w)$, we obtain four independent recursive equations for the axial spectra $\hat{V}_{+}(w)$, $\hat{I}_{+}(w)$, $\hat{V}_{\pi+}(w_1)$, $\hat{I}_{\pi+}(w_1)$. After algebraic manipulations, these equations can be rewritten as:

$$\hat{V}_d(w) = \frac{Z_1 \sin(w+\Phi) - Z_o \sin(g(w+\Phi))}{Z_1 \sin(w+\Phi) + Z_o \sin(g(w+\Phi))} \hat{V}_d(w+2\Phi) + \frac{2Z_o \sin(w+\Phi)}{\left[Z_1 \sin(w+\Phi) + Z_o \sin(g(w+\Phi))\right]} \hat{V}_{\pi d}(g(w+\Phi)+\Phi_1) \tag{6.4.13}$$

$$\hat{V}_{\pi d}(w_1) = \frac{2Z_1 \sin(w_1+\Phi_1)}{\left[Z_1 \sin(g_1(w_1+\Phi_1) + Z_o \sin(w_1+\Phi_1)\right]} \hat{V}_d(g_1(w_1+\Phi_1)+\Phi) - \frac{Z_1 \sin(g_1(w_1+\Phi_1)) - Z_o \sin(w_1+\Phi_1)}{Z_1 \sin(g_1(w_1+\Phi_1)) + Z_o \sin(w_1+\Phi_1)} \hat{V}_{\pi d}(w_1+2\Phi_1) \tag{6.4.14}$$

$$\hat{I}_+(w) = -\frac{Z_1 \sin(w+\Phi) - Z_o \sin(g(w+\Phi))}{Z_1 \sin(w-\Phi) + Z_o \sin(g(w-\Phi))} \hat{I}_+(w+2\Phi) + \frac{2Z_1 \sin(w+\Phi)}{Z_1 \sin(w+\Phi) + Z_o \sin(g(w+\Phi))} \hat{I}_{\pi+}(g(w+\Phi)+\Phi_1) \tag{6.4.15}$$

$$\hat{I}_{\pi+}(w_1) = \frac{2Z_o \sin(w_1-\Phi_1)}{Z_1 \sin(g_1(w_1-\Phi_1)) + Z_o \sin(w_1-\Phi_1)} \hat{I}_+(g_1(w_1+\Phi_1)+\Phi) + \frac{Z_1 \sin(g_1(w_1+\Phi_1)) - Z_o \sin(w_1+\Phi_1)}{Z_1 \sin(g_1(w_1+\Phi_1)) + Z_o \sin(w_1+\Phi_1)} \hat{I}_{\pi+}(w_1+2\Phi_1) \tag{6.4.16}$$

where we have introduced the odd functions $\hat{V}_d(w)$ and $\hat{V}_{\pi d}(w_1)$ defined by:

$$\hat{V}_d(w) = \sin(w)\hat{V}_+(w), \quad \hat{V}_{\pi d}(w_1) = \sin(w_1)\hat{V}_{\pi+}(w_1) \tag{6.4.17}$$

6.5. Some remarks on the Fredholm integral equations (6.3.24), (6.3.26) and numerical solutions

The previous sections showed that the IEs (6.3.22)–(6.3.28) for the solution on the problem, the representations (6.3.29)–(6.3.32) and the analytic continuations (6.4.13)–(6.4.16) provide the axial spectra $\hat{V}_d(w)$, $\hat{V}_{\pi d}(w_1)$, $\hat{I}_{\pi+}(w_1)$, $\hat{I}_+(w)$ of the problem.

To infer the properties of the axial spectra, it is interesting to note that for the normalized values $\underline{\eta} = \eta / k_o$, $\underline{m} = m / k_o$, $\underline{m}_1 = m_1 / k_o$, the normalized unknowns $\underline{Y}_i(\underline{\eta}) = k_o\, Y_i(\eta)$, $\underline{V}_{a,b+}(-\underline{m}) = k_o\, V_{a,b+}(-m)$, $\underline{I}_{a,b+}(-\underline{m}) = k_o\, I_{a,b+}(-m)$, $\underline{V}_{a,b+}(-\underline{m}_1) = k_o\, V_{a,b+}(-m_1)$ and $\underline{I}_{a,b+}(-\underline{m}_1) = k_o\, I_{a,b+}(-m_1)$ satisfy the GWHEs of our problem.

This means that, if we obtain a solution $F(\eta)$ assuming a given arbitrary value of k, the function $F_o(\eta) = \frac{k}{k_x} F(\frac{k}{k_x}\eta)$ represents the solution when the propagation constant assumes the different value k_x. At the same time, if $f(\rho)$ is the inverse Laplace transform relevant to k, the function $f(\frac{k_x}{k}\rho)$ is the inverse Laplace transform relevant to k_x. This fact allows us to assume suitable values of k for the numerical solutions for the FIEs.

In Appendix C of Daniele (2011b), it has been proved that the kernels of equations (6.3.24) and (6.3.26) are representative of a compact operator. Hence, the Fredholm factorization in principle provides an exact solution of the dielectric wedge problem by using the method of the determinants. As is well known, this solution is impracticable because it requires the computation of multiple integrals of increasing order. Therefore, in practice, it is better to use the very efficient approximate methods of the Fredholm integral equation of the second kind, as suggested in previous chapters and in particular in Chapter 5. Furthermore, in this case, we obtain very accurate numerical results by using a simple quadrature of the occurring integrals. In order to simplify the numerical procedure, we consider only acute dielectric wedges that yield aperture angle $\Phi \geq \frac{\pi}{2}$.

It is important to obtain the asymptotic behavior of the functions $P_{2,4}(u)$ and $Q_{2,4}(v)$ as $u \to \infty$ and $v \to \infty$. Taking into account that $I_{\rho+}(\eta,\varphi) = O\left[\eta^{-c}\right]$, $\mathrm{Re}[c] > 0$, from definitions (6.3.9), we obtain:

$$P_2(u) = e^{-|u|} O\left[\exp(-c\frac{\Phi}{\pi}|u|)\right], \; P_4(u) = e^{-|u|/2} O\left[\exp(-c\frac{\Phi}{\pi}|u|)\right] \tag{6.5.1}$$

In practice, $(e^{|u|}P_2(u), e^{|u|/2}P_4(u))$ are vanishing for $u > 3\frac{\pi}{c\Phi}$. The parameter c can be determined by studying the static edge field behavior. For the dielectric wedge, we have (Scharstein 1993).

$$\sin(c\pi) + \frac{\varepsilon_r - 1}{\varepsilon_r + 1}\sin[c(2\Phi_1 - \pi)] = 0 \tag{6.5.2}$$

Since the minimum value holds for the PEC wedge ($c_{PEC} = \frac{\pi}{2\Phi}$), it follows that $\mathrm{Re}[c] > c_{PEC}$.

In the quadrature scheme, the intervals of integrals are truncated. We can have considerable relative errors for high values of $\mathrm{Im}[w]$. These errors do not influence the far field since its computation requires the accurate evaluation of the spectra only for real values of w. Conversely, the errors are important for the near field, since Meixner's condition requires $\mathrm{Im}[w] \to \infty$.

However, in practice, we can ignore these conditions, and in the η plane, a band $|\eta / k_o| \approx 10 \to \mathrm{Im}[w] \approx 3$ could also be sufficient for an acceptable evaluation of the near field (Daniele 2011a).

A property of the kernel $M(u,u') = \frac{1}{\pi j}\frac{e^u}{(e^u - j)}\frac{e^{u'} + j}{e^{u+u'} + 1}$ is that it satisfies the condition

$$M(u,u') = M^*(-u,-u') \tag{6.5.3}$$

where * denotes complex conjugation.

Consequently, by considering the complex conjugate of equations (6.3.24) and (6.3.26), we obtain that $-P_i^*(-u)$, $-Q_i^*(-v)$ (i = 1,2,3,4) are solutions of the problem too.

Since the solution is unique, it yields:

$$P_i(u) = -P_i^*(-u) \quad Q_i(v) = -Q_i^*(-u) \tag{6.5.4}$$

Taking into account Schwartz's reflection principle, it follows that

$$jk_o\hat{V}_+(w) = [jk_o\hat{V}_+(w^*)]^* \tag{6.5.5}$$

The numerical solutions of (6.3.24) and (6.3.26) confirmed the validity of equation (6.5.4).

As a result of (6.5.4), the starting spectra evaluated with (6.3.29)–(6.3.30) (or (6.3.31)–(6.3.32)) are purely imaginary in the regularity strips $-\Phi < \mathrm{Re}[w] < 0$ (or $-\Phi_1 < \mathrm{Re}[w_1] < 0$).

However, considering the analytic continuation of the starting spectra (6.4.13)–(6.4.16), the above properties do no longer hold for spectra defined out the regularity strip.

The integral equations can be rewritten on the half-line $0 \le u < \infty$ rather than on the whole line $-\infty < u < \infty$. The equations on the half-line can be easily obtained assuming the real and imaginary parts of (6.3.24) and (6.3.26) and by taking into account the properties $\mathrm{Re}[P_i(-u)] = -\mathrm{Re}[P_i(u)]$, $\mathrm{Im}[P_i(-u)] = \mathrm{Im}[P_i(u)]$ $(i = 1,..,4)$.

As we said, since the kernels of the integral equations (6.3.24) and (6.3.26) present a well-suited behavior, we use a simple sample-and-hold quadrature scheme to obtain accurate and stable numerical solutions. We apply uniform sampling $f(hi)$ with $i = -A/h......A/h$ and modified left-rectangle numerical integration formula

$$\int_{-\infty}^{\infty} f(u)du \approx h \sum_{i=-A/h}^{A/h} f(hi) \tag{6.5.6}$$

where A and h are respectively the truncation parameter and the step parameter for the integrals in u.

This rule has been successfully applied for the impenetrable wedge case (Chapter 5). The total number of samples is $2A/h+1$. We note that as $A \to +\infty$ and $h \to 0$, the numerical solution of the Fredholm integral equation converges to the exact solution (Kantorovich and Krylov 1964); consequently, h has to be chosen as small as possible and A has to be chosen as large as possible with the compromise on the conditioning of the resulting linear system.

The selection of A and h can be distinguished for the exterior problem from the interior problem respectively modeled by the first equations of (6.3.24) and (6.3.26) and the second equations of (6.3.24) and (6.3.26).

However, according to our experience, we note that $A = A_1 = 10$ and $h = h_1 = 0.05$ (A_1 and h_1 are related to the second equations in (6.3.24) and (6.3.26)) yield stable and accurate solutions in terms of $P_i(u)$ and $Q_i(u)$ samples, voltages' and currents' spectra and thus field components; see previous section.

Finally, it is interesting to note that, for selected values of the aperture angle of the wedge (e.g. for right wedges), the kernels $M_{2e}(u,u')$ and $M_{4e}(u,u')$ that are expressed by integrals can be evaluated in closed form.

6.6. Field evaluation in any point of the space

In the first five sections, the scope has been the evaluation of the axial spectra (the plus unknowns of the GWHEs). However, the real scope is the evaluation of the spectra for every direction φ in order to obtain field estimation in any point of the space. For this scope, we apply the rotating wave method (Chapter 3).

To make the evaluation symmetric in free space and in the wedge, it is convenient to introduce the angular coordinate defined by $\varphi_1 = \pi - \varphi$ inside the wedge. In this way, we have for free space $0 < |\varphi| < \Phi$, while for the dielectric wedge $0 < |\varphi_1| \leq \Phi_1$.

Given the axial spectra $\hat{V}_d(w)$, $\hat{I}_+(w)$, $\hat{V}_{\pi d}(w_1)$ and $\hat{I}_{\pi +}(w_1)$, (3.13.8)–(3.13.9) provide the rotating waves in the air and in the dielectric. Hence, using equations (3.11.8), (3.11.10), (3.11.13) and (3.11.14), the following representations of the spectra for an arbitrary direction φ hold (Daniele 2003a, 2011a; Daniele and Lombardi 2011):

$$V_{\varphi d}(w) = \frac{Z_o\left[\hat{I}_+(w-\varphi) - \hat{I}_+(w+\varphi)\right] + \hat{V}_d(w-\varphi) + \hat{V}_d(w+\varphi)}{2} \quad (6.6.1)$$

$$I_{\varphi +}(w) = \frac{Z_o\left[\hat{I}_+(w-\varphi) + \hat{I}_+(w+\varphi)\right] + \hat{V}_d(w-\varphi) - \hat{V}_d(w+\varphi)}{2Z_o} \quad (6.6.2)$$

for $0 < |\varphi| < \Phi$ (we have omitted "o" in the subscripts of axial spectra) and

$$V_{\varphi_1 d}(w_1) = \frac{Z_1\left[\hat{I}_{\pi +}(w_1+\varphi_1) - \hat{I}_{\pi +}(w_1-\varphi_1)\right] + \hat{V}_{\pi d}(w_1+\varphi_1) + \hat{V}_{\pi d}(w_1-\varphi_1)}{2} \quad (6.6.3)$$

$$I_{\varphi_1 +}(w_1) = \frac{Z_1\left[\hat{I}_+(w_1+\varphi_1) + \hat{I}_+(w_1-\varphi_1)\right] + \hat{V}_{\pi d}(w_1+\varphi_1) - \hat{V}_{\pi d}(w_1-\varphi_1)}{2Z_1} \quad (6.6.4)$$

for $0 < |\varphi_1| < \Phi_1$

where $V_{\varphi d}(w) = \sin(w) V_{z+}(-k\cos(w), \varphi)$, $V_{\varphi_1 d}(w_1) = \sin(w_1) V_{z+}(-k_1\cos(w_1), \varphi_1)$, $I_{\varphi +}(w) = I_{\rho +}(-k\cos(w), \varphi)$, $I_{\varphi_1 +}(w_1) = I_{\rho +}(-k_1\cos(w_1), \varphi_1)$.

In the literature, to express the field at any direction φ in the w plane, the Sommerfeld function is used. Taking into account the link between the Sommerfeld function and the rotating wave $v_1(w)$ (3.12.17) and the link between the rotating waves and the spectra (see Chapter 3), we obtain the following expressions of the Sommerfeld function in terms of the axial spectra:

$$s_E(w) = \frac{jk_o}{2}[-\sin w\hat{V}_+(w) + Z_o\hat{I}_+(w)] \text{ in air} \quad (6.6.5)$$

$$s_E^{\text{int}}(w_1) = \frac{jk_o}{2}[-\sin w_1\hat{V}_{\pi +}(w_1) + Z_{o1}\hat{I}_{\pi +}(w_1)] \text{ in the dielectric} \quad (6.6.6)$$

Following the procedure presented in section 5.7 for the impenetrable wedge, we obtain the total field in terms of GO and UTD components for the dielectric wedge by distinguishing the exterior from the interior problem. For the exterior problem, we have a direct application of the equations of Chapter 5, while for the interior, we need some modifications of the equations in terms of the reference system and physical parameters of the interior problem (see section IV of Daniele and Lombardi (2011) for further details). For example, the GTD field components of the exterior and interior problems assume the forms:

$$E_z^{gtd}(\rho,\varphi) = E_o \frac{e^{-j(k_o\rho+(\pi/4))}}{\sqrt{2\pi k_o \rho}} D_E^{ext}(\varphi,\varphi_o), \quad -\Phi<\varphi<\Phi \tag{6.6.7}$$

with

$$D_E^{ext}(\varphi,\varphi_o) = \frac{s_E(\varphi-\pi)-s_E(\varphi-\pi)}{E_o} \tag{6.6.8}$$

and

$$E_z^{gtd}(\rho,\varphi_1) = E_o \frac{e^{-j(k_1\rho+(\pi/4))}}{\sqrt{2\pi k_1 \rho}} D_E^{\text{int}}(\varphi_1,\varphi_o), \quad -\Phi_1<\varphi_1<\Phi_1 \tag{6.6.9}$$

with

$$D_E^{\text{int}}(\varphi_1,\varphi_o) = \frac{s_E^{\text{int}}(-\varphi_1-\pi)-s_E^{\text{int}}(-\varphi_1-\pi)}{E_o} \tag{6.6.10}$$

We note from (6.6.1)–(6.6.10) that the computation of GTD components requires the knowledge of the outer axial spectra in the range $-\pi-\Phi<w<-\pi+\Phi$ and the knowledge of the inner axial spectra in the range $-\pi-\Phi_1<w_1<-\pi+\Phi_1$.

Uniformization of (6.6.7) and (6.6.10) is obtained as in section 5.7 for the impenetrable wedge problem.

The far-field evaluation for selected wedges and incident angles as well as the numerical validation in test cases are presented in sections 6.8 and 6.9 as well as in section V of Daniele and Lombardi (2011). The numerical results show the validity and efficiency of the Fredholm factorization method to solve dielectric (penetrable) wedge problems using the WH technique.

The WH solution of the dielectric wedge problem presented in this chapter is the first one that the authors of this book have developed in the literature. It contains very sophisticated mathematical methods to overcome difficult analytical problems related to the presence of two media. Recently, many improvements on the WH method have been made to study the diffraction in the presence of several wedges and layers (Daniele and Lombardi 2016; Daniele *et al.* 2017a, 2017b, 2017c, 2018, 2019). In particular, simple circuit models allow us to quickly formulate the Fredholm factorization methods of these problems. Perhaps these improvements will allow us to obtain a more conceptually simple solution of the one presented here.

6.7. The dielectric wedge at skew incidence

The solution procedure presented for the scattering of a dielectric wedge at normal incidence (E_z-polarization) is extended (Daniele 2011b; Daniele and Lombardi 2011) in the presence of an incident plane wave at skew incidence (Figure 6.1.1)

$$E_z^i = E_o e^{jk_o \rho\cos(\varphi-\varphi_o)} e^{-j\alpha_o z}, \quad H_z^i = H_o e^{j\tau_o \rho\cos(\phi-\phi_o)} e^{-j\alpha_o z} \tag{6.7.1}$$

where E_o and H_o are known quantities, β is the angle between the incident direction $\hat{\mathbf{n}}_i = -\hat{x}\sin\beta\cos\varphi_o - \hat{y}\sin\beta\sin\varphi_o + \hat{z}\cos\beta$ and $\hat{z}$, $k = k_o = \omega\sqrt{\mu_o\varepsilon_o}$, $\alpha_o = k_o\cos\beta$ and $\tau_o = k\sin\beta = \sqrt{k^2-\alpha_o^2}$.

The skew incident plane wave introduces new parameters and functions defined in the following equations:

$$\alpha_1 = k_1\sin\beta_1 = \alpha_o, \; \eta_1 = \eta_o, \; \varepsilon_1 = \varepsilon_r\varepsilon_o, \; \mu_1 = \mu_r\mu_o, \; k_1 = \omega\sqrt{\mu_1\varepsilon_1} = \sqrt{\varepsilon_r\mu_r}\,k_o,$$

$$Z_o = \sqrt{\frac{\mu_o}{\varepsilon_o}}, \; Z_1 = \sqrt{\frac{\mu_o}{\varepsilon_r\varepsilon_o}}$$

$$\tau_o = \sqrt{k_o^2-\alpha_o^2}, \; \tau_1 = \sqrt{k_1^2-\alpha_1^2} = \lambda_e\tau_o, \; \lambda_e = \sqrt{1+(\varepsilon_r\mu_r-1)\csc^2\beta}$$

$\xi = \xi(\eta) = \sqrt{\tau_o^2-\eta^2}$, $\quad \xi_1 = \xi_1(\eta) = \sqrt{\tau_1^2-\eta_1^2}$ with the branch $\xi(0) = \tau_o$, $\xi_1(0) = \tau_1$, $n_1 = n_1(\eta_1) = -\xi_1\cos\Phi_1 - \eta_1\sin\Phi_1$

$$m = m(\eta) = -\eta\cos\Phi + \xi\sin\Phi, \; n = n(\eta) = -\xi\cos\Phi - \eta\sin\Phi$$

$$m_1 = m_1(\eta_1) = -\eta_1\cos\Phi_1 + \xi_1\sin\Phi_1,$$

As reported in Daniele (2011b) and Daniele and Lombardi (2011), complex deductions, which are conceptually similar to those of the normal incidence case, yield two uncoupled similar systems of FIE of order 4 having as unknowns the eight auxiliary spectra $P_i(u)$ (i = 1,2…,8)

$$\begin{cases} P_1(u) - \tan\dfrac{-\frac{\pi}{2} + ju}{2} P_2(u) - \int_{-\infty}^{\infty} M(u,u')P_2(u')du' = n_1(u) \\ P_7(u) - \tan\dfrac{-\frac{\pi}{2} + ju}{2} P_8(u) - \int_{-\infty}^{\infty} M(u,u')P_8(u')du' = n_7(u) \\ h_{ae11}(u)\,P_1(u) + \tan\dfrac{-\frac{\pi}{2} + jv(u)}{2} h_{ae33}(u)\,P_2(u) + \int_{-\infty}^{\infty} M_{ae32}(u,u')P_7(u')du' \\ \qquad + \int_{-\infty}^{\infty} M_{ae33}(u,u')P_2(u')du' = 0 \\ h_{ae22}(u)\,P_7(u) + \tan\dfrac{-\frac{\pi}{2} + jv(u)}{2} h_{ae44}(u)\,P_8(u) + \int_{-\infty}^{\infty} M_{ae41}(u,u')P_1(u')du' \\ \qquad + \int_{-\infty}^{\infty} M_{ae44}(u,u')P_8(u')du' = 0 \end{cases} \tag{6.7.2}$$

$$\begin{cases} P_3(u) - \tan\dfrac{-\frac{\pi}{2} + ju}{2} P_4(u) - \int_{-\infty}^{\infty} M(u,u')P_4(u')du' = n_3(u) \\ P_5(u) - \tan\dfrac{-\frac{\pi}{2} + ju}{2} P_6(u) - \int_{-\infty}^{\infty} M(u,u')P_6(u')du' = n_5(u) \\ h_{be11}(u)\,P_3(u) + \tan\dfrac{-\frac{\pi}{2} + jv(u)}{2} h_{be33}(u)\,P_4(u) + \int_{-\infty}^{\infty} M_{be32}(u,u')P_5(u')du' \\ \qquad + \int_{-\infty}^{\infty} M_{be33}(u,u')P_4(u')du' = 0 \\ h_{be22}(u)\,P_5(u) + \tan\dfrac{-\frac{\pi}{2} + jv(u)}{2} h_{be44}(u)\,P_6(u) + \int_{-\infty}^{\infty} M_{be41}(u,u')P_3(u')du' \\ \qquad + \int_{-\infty}^{\infty} M_{be44}(u,u')P_6(u')du' = 0 \end{cases} \tag{6.7.3}$$

with

$$v = v(u) = \frac{j\pi}{2\Phi_1}\left(\Phi_1 - 2\arccos\frac{\cos\frac{(\pi + 2ju)\Phi}{2\pi}}{\sqrt{\varepsilon_{tr}}}\right) \tag{6.7.4}$$

In the next equations we assume

$$i, j = 1,2,3,4 \tag{6.7.5}$$

$$M(u,u') = \frac{1}{\pi j}\frac{e^u}{\left(e^u - j\right)}\frac{e^{u'} + j}{e^{u+u'} + 1}, \tag{6.7.6}$$

$$M_{aei\,j}(u,u') = \int_{-\infty}^{\infty} M(v(u),v')T_{aei\,j}(v',u')dv', \tag{6.7.7}$$

$$M_{bei\,j}(u,u') = \int_{-\infty}^{\infty} M(v(u),v')T_{bei\,j}(v',u')dv' \tag{6.7.8}$$

$$T_{ae11} = T(v,u)\cosh(u)\csc\frac{(\pi + 2ju)\Phi}{2\pi}sech(v)\sin\frac{(\pi + 2jv)\Phi_1}{2\pi} \tag{6.7.9}$$

$$T_{ae22} = T(v,u)\cos\frac{1}{4}(\pi - 2ju)\csc\frac{(\pi + 2ju)\Phi}{2\pi}Sec\frac{1}{4}(\pi - 2jv)\sin\frac{(\pi + 2jv)\Phi_1}{2\pi} \tag{6.7.10}$$

$$T_{ae32} = T(v,u)Z_1 \cos\frac{1}{4}(\pi - 2ju)\cos(\beta)[\varepsilon_{tr} \cot\frac{(\pi + 2ju)\Phi}{2\pi} - \cos\frac{(\pi + 2jv)\Phi_1}{2\pi}\csc\frac{(\pi + 2ju)\Phi}{2\pi}]\csc^2\frac{(\pi - 2jv)}{4} \tag{6.7.11}$$

$$T_{ae33} = T(v,u)\frac{Z_1\varepsilon_{tr}(j+\sinh u)}{Z_o(j+\sinh v)} \tag{6.7.12}$$

$$T_{ae41} = T(v,u)\cos(\beta)\cosh(u)\frac{[\varepsilon_{tr}\cot\frac{(\pi+2ju)\Phi}{2\pi} - \cos\frac{(\pi+2jv)\Phi_1}{2\pi}\csc\frac{(\pi+2ju)\Phi}{2\pi}]\csc\frac{(\pi-2jv)}{4}}{2Z_o} \tag{6.7.13}$$

$$T_{ae44} = T(v,u)\varepsilon_{tr}\csc\frac{(\pi-2jv)}{4}\sin\frac{(\pi-2ju)}{4} \tag{6.7.14}$$

$$T_{be22} = T_{ae11},\quad T_{be11} = -T_{ae22} \tag{6.7.15}$$

$$T_{be32} = Z_1 Z_o T_{ae41},\quad T_{be33} = \frac{Z_1}{Z_o}T_{ae44} \tag{6.7.16}$$

$$T_{be41} = -\frac{1}{Z_1 Z_o}T_{ae32} \tag{6.7.17}$$

$$T_{be44} = \frac{Z_o}{Z_1}T_{ae33} \tag{6.7.18}$$

$$T(v,u) = -\frac{\Phi}{2\pi^2}\frac{\sin[\frac{\Phi}{\pi}(ju+\frac{\pi}{2})+\Phi]}{\cos[\frac{\Phi}{\pi}(ju+\frac{\pi}{2})+\Phi]-\sqrt{\varepsilon_{tr}}\cos[\frac{\Phi_1}{\pi}(jv+\frac{\pi}{2})+\Phi_1]} \tag{6.7.19}$$

$$h_{ae11} = \cosh(u)\csc\frac{(\pi+2ju)\Phi}{2\pi}sech(v(u))\sin\frac{(\pi+2jv(u))\Phi_1}{2\pi} \tag{6.7.20}$$

$$h_{ae22} = \cos\frac{1}{4}(\pi-2ju)\csc\frac{(\pi+2ju)\Phi}{2\pi}sec\frac{1}{4}(\pi-2jv(u)) \sin\frac{(\pi+2jv(u))\Phi_1}{2\pi} \tag{6.7.21}$$

$$h_{ae32} = Z_1 \cos\frac{1}{4}(\pi - 2ju)\cos(\beta)[\varepsilon_{tr} \cot\frac{(\pi + 2ju)\Phi}{2\pi} - \cos\frac{(\pi + 2jv(u))\Phi_1}{2\pi} \csc\frac{(\pi + 2ju)\Phi}{2\pi}] \csc^2 \frac{(\pi - 2jv(u))}{4} \tag{6.7.22}$$

$$h_{ae33} = \frac{Z_1 \varepsilon_{tr}(j + \sinh u)}{Z_o(j + \sinh v(u))} \tag{6.7.23}$$

$$h_{ae41} = \cos(\beta)\cosh(u)\frac{[\varepsilon_{tr}\cot\frac{(\pi + 2ju)\Phi}{2\pi} - \cos\frac{(\pi + 2jv(u))\Phi_1}{2\pi}\csc\frac{(\pi + 2ju)\Phi}{2\pi}]\csc\frac{(\pi - 2jv(u))}{4}}{2Z_o} \tag{6.7.24}$$

$$h_{ae44} = \varepsilon_{tr}\csc\frac{(\pi - 2jv(u))}{4}\sin\frac{(\pi - 2ju)}{4} \tag{6.7.25}$$

$$h_{be22} = h_{ae11}, \quad h_{be11} = -h_{ae22} \tag{6.7.26}$$

$$h_{be32} = Z_1 Z_o h_{ae41}, \quad h_{be33} = \frac{Z_1}{Z_o} h_{ae44} \tag{6.7.27}$$

$$h_{be41} = -\frac{1}{Z_1 Z_o} h_{ae32} \tag{6.7.28}$$

$$h_{be44} = \frac{Z_o}{Z_1} h_{ae33} \tag{6.7.29}$$

The sources $n_j(u)$, j=1,3,5,7 are described here in the case $\varphi_o < \frac{\Phi}{2}$.

The general case is considered in Daniele (2011b):

$$n_1(u)=-\frac{-j4E_o\frac{\pi}{\Phi_i}}{j\sinh u-\cos\frac{\pi}{\Phi}\varphi_o},n_3(u)=-\frac{j4E_o\frac{\pi}{\Phi}\sin(\frac{\pi\varphi_o}{2\Phi})}{j\sinh u-\cos\frac{\pi}{\Phi}\varphi_o},$$
$$n_5(u)=-\frac{j4H_o\frac{\pi}{\Phi}}{j\sinh u-\cos\frac{\pi}{\Phi}\varphi_o},n_7(u)=-\frac{-j4H_o\frac{\pi}{\Phi}\sin(\frac{\pi\varphi_o}{2\Phi})}{j\sinh u-\cos\frac{\pi}{\Phi}\varphi_o}\tag{6.7.30}$$

Related to $P_i(u)$, we have eight other auxiliary spectra $Q_i(u)$ (i = 1,2…,8) defined by:

$$\begin{vmatrix}Q_1(v)\\Q_7(v)\\Q_2(v)\\Q_8(v)\end{vmatrix}=\int_{-\infty}^{\infty}T_{ae}(v,u)\begin{vmatrix}P_1(u)\\P_7(u)\\P_2(u)\\P_8(u))\end{vmatrix}du,\quad\begin{vmatrix}Q_3(v)\\Q_5(v)\\Q_4(v)\\Q_6(v)\end{vmatrix}=\int_{-\infty}^{\infty}T_{be}(v,u)\begin{vmatrix}P_3(u)\\P_5(u)\\P_4(u)\\P_6(u))\end{vmatrix}du\tag{6.7.31}$$

where the matrices $T_{ae}(v,u)$ and $T_{be}(v,u)$ are given by

$$T_{ae}(v,u)=\begin{vmatrix}T_{ae11}&0&0&0\\0&T_{ae22}&0&0\\0&T_{ae32}&T_{ae33}&0\\T_{ae41}&0&0&T_{ae44}\end{vmatrix},T_{be}(v,u)=\begin{vmatrix}T_{be11}&0&0&0\\0&T_{be22}&0&0\\0&T_{be32}&T_{be33}&0\\T_{be41}&0&0&T_{be44}\end{vmatrix}\tag{6.7.32}$$

The auxiliary spectra $P_i(u)$, $Q_i(u)$ (i = 1,..8) define the functions $Y_i(-\tau_o\cos w)$ given for $i=1,3,5,7$ by:

$$Y_i(-\tau_o\cos w)=\frac{1}{\tau_o}\left[\int_{-\infty}^{\infty}M(u,u')P_{i+1}(u')du'+n_i(u)\right]\tag{6.7.33}$$

$$Y_{i+1}(-\tau_1\cos w_1)=\frac{1}{\tau_o}\int_{-\infty}^{\infty}M(v,v')Q_{i+1}(v')dv'\tag{6.7.34}$$

Finally, after many algebraic manipulations (Daniele 2011b), we obtain the following representations of the Laplace transforms of the axial electromagnetic field valuated in the planes w and w_1 (plus axial spectra):

$$V_{\rho+}(-\tau_o \cos w, 0) = \hat{V}_{\rho+}(w) = -\frac{1}{2}\cos\frac{\pi w}{2\Phi}\left(\cot w \cot\beta \sin\frac{\pi w}{2\Phi} Y_1(w) + Z_o \csc\beta\, Y_7(w)\right)$$

$$V_{z+}(-\tau_o \cos w, 0) = \hat{V}_{z+}(w) = -\frac{1}{4}\csc w \sin\frac{\pi w}{\Phi} Y_1(w)$$

$$\hat{I}_{\rho+}(w) = \frac{1}{2Z_o}\cos\frac{\pi w}{2\Phi}\left(-\csc\beta\, Y_3(w) + Z_o \cot w \cot\beta \sin\frac{\pi w}{2\Phi} Y_5(w)\right)$$

$$I_{z+}(-\tau_o \cos w, 0) = \hat{I}_{z+}(w) = \frac{1}{4}\csc w \sin\frac{\pi w}{\Phi} Y_5(w)$$

$$V_{\rho+}(-\tau_1 \cos w_1, \pi) = \hat{V}_{\rho\pi+}(w_1) = \frac{1}{2\lambda_e}\cos\frac{\pi w_1}{2\Phi_1}\left(-\cot w_1 \cot\beta \sin\frac{\pi w_1}{2\Phi_1} Y_2(w_1) + Z_o \csc\beta\, Y_8(w_1)\right)$$

$$V_{z+}(-\tau_1 \cos w_1, \pi) = \hat{V}_{z\pi+}(w_1) = -\frac{1}{4}\csc w_1 \sin\frac{\pi w_1}{\Phi_1} Y_2(w_1)$$

$$I_{\rho+}(-\tau_1 \cos w_1, \pi) = \hat{I}_{\rho\pi+}(w_1) = \frac{1}{2Z_o\lambda_e}\cos\frac{\pi w_1}{2\Phi_1}\left(\varepsilon_r \csc\beta Y_4(w_1) + Z_o \cot w_1 \cot\beta \sin\frac{\pi w_1}{2\Phi_1} Y_6(w_1)\right)$$

$$I_{z+}(-\tau_1 \cos w_1, \pi) = \hat{I}_{z\pi+}(w_1) = \frac{1}{4}\csc w_1 \sin\frac{\pi w_1}{\Phi_1} Y_6(w_1) \qquad (6.7.35)$$

These spectra have validity strips $-\Phi \le \mathrm{Re}[w] \le \Phi$ or $-\Phi_1 \le \mathrm{Re}[w_1] \le \Phi_1$ as they are respectively defined in free space or in dielectric wedge.

By repeating the procedure illustrated at normal incidence, after cumbersome algebraic manipulations, we obtain the recursive equations

$$\begin{vmatrix} \hat{V}_{\rho+}(w) \\ \hat{V}_{z+}(w) \\ \hat{I}_{z+}(w) \\ \hat{I}_{\rho+}(w) \end{vmatrix} = H_{11n}[w+\Phi, g_t(w+\Phi)] \begin{vmatrix} \hat{V}_{\rho+}(w+2\Phi) \\ \hat{V}_{z+}(w+2\Phi) \\ \hat{I}_{z+}(w+2\Phi) \\ \hat{I}_{\rho+}(w+2\Phi) \end{vmatrix} + H_{12n}[w+\Phi, g_t(w+\Phi)] \begin{vmatrix} \hat{V}_{\rho\pi+}(g_t(w+\Phi)+\Phi_1) \\ \hat{V}_{z\pi+}(g_t(w+\Phi)+\Phi_1) \\ \hat{I}_{z\pi+}(g_t(w+\Phi)+\Phi_1) \\ \hat{I}_{\rho\pi+}(g_t(w+\Phi)+\Phi_1) \end{vmatrix} \tag{6.7.36}$$

$$\begin{vmatrix} \hat{V}_{\rho\pi+}(w_1) \\ \hat{V}_{z\pi+}(w_1) \\ \hat{I}_{z\pi+}(w_1) \\ \hat{I}_{\rho\pi+}(w_1) \end{vmatrix} = H_{21n}[g_{t1}(w_1+\Phi_1), w_1+\Phi_1] \begin{vmatrix} \hat{V}_{\rho+}(g_{t1}(w_1+\Phi_1)+\Phi) \\ \hat{V}_{z+}(g_{t1}(w_1+\Phi_1)+\Phi) \\ \hat{I}_{z+}(g_{t1}(w_1+\Phi_1)+\Phi) \\ \hat{I}_{\rho+}(g_{t1}(w_1+\Phi_1)+\Phi) \end{vmatrix} + H_{22n}[g_{t1}(w_1+\Phi_1), w_1+\Phi_1] \begin{vmatrix} \hat{V}_{\rho\pi+}(w_1+2\Phi_1) \\ \hat{V}_{z\pi+}(w_1+2\Phi_1) \\ \hat{I}_{z\pi+}(w_1+2\Phi_1) \\ \hat{I}_{\rho\pi+}(w_1+2\Phi_1) \end{vmatrix} \tag{6.7.37}$$

where $w_1 = g_e(w), \quad w = g_{1e}(w_1)$ with

$$g_e(w) = -\arccos\left(\frac{\cos w}{\lambda_e}\right), \quad g_1(w_1) = -\arccos(\lambda_e \cos w_1) \tag{6.7.38}$$

The matrix $H(w,w_1) = \begin{vmatrix} H_{11n}(w,w_1) & H_{12n}(w,w_1) \\ H_{21n}(w,w_1) & H_{22n}(w,w_1) \end{vmatrix}$ requires the use of the modern technical computing software. The expression of the matrix $H(w,w_1)$ is very long, and it is given in the Appendix of Daniele (2009).

Given the whole axial spectra, we can obtain the spectra for any direction φ using the rotating wave method (Chapter 3). In the air, we obtain the following expressions of the spectra for any direction $-\Phi \le \varphi \le \Phi$:

$$V_{z\varphi+}(w) = \frac{v_1(w+\varphi)+v_2(w-\varphi)}{-\tau_o \sin w}, \quad I_{z\varphi+}(w) = \frac{i_1(w+\varphi)+i_2(w-\varphi)}{-\tau_o \sin w} \tag{6.7.39}$$

where the rotating waves are expressed in terms of the axial spectra $\hat{V}_{z+}(w)$, $\hat{I}_{z+}(w)$ by the following equations (3.3.8), (3.3.9):

$$\begin{aligned} v_1(w) = &-\frac{1}{2}k_o \sin\beta(Z_o \cos w \cos\beta \hat{I}_{z+}(w) \\ &- Z_o \sin\beta \hat{I}_{\rho+}(w) + \sin(w)\hat{V}_{z+}(w)) \end{aligned} \tag{6.7.40}$$

$$i_1(w) = -\frac{k_o \sin\beta(Z_o \sin w\, \hat{I}_{z+}(w) - \cos\beta \cos w \hat{V}_{z+}(w) + \sin(\beta)\hat{V}_{\rho+}(w))}{2Z_o} \tag{6.7.41}$$

with the property $v_2(w) = -v_1(-w), \; i_2(w) = -i_1(-w)$.

Similarly in the wedge $(-\Phi_1 \le \varphi_1 \le \Phi_1, \; \varphi_1 = \pi - \varphi)$, we obtain:

$$V_{z\varphi d+}(w_1) = \frac{v_{1d}(w_1+\varphi)+v_{2d}(w_1-\varphi)}{-\tau_1 \sin w_1}, I_{z\varphi d+}(w_1) = \frac{i_{1d}(w_1+\varphi)+i_{2d}(w_1-\varphi)}{-\tau_1 \sin w_1} \tag{6.7.42}$$

where the rotating waves are expressed in terms of the axial spectra $\hat{V}_{z\pi+}(w)$, $\hat{I}_{z\pi+}(w)$ by the following equations:

$$v_{d1}(w) = -\frac{1}{2}k_1 \sin\beta_1(-Z_1 \cos w_1 \cos\beta_1 \hat{I}_{z\pi+}(-\pi + w_1) - Z_1 \sin\beta_1 \hat{I}_{\rho\pi+}(-\pi + w_1) - \sin(w_1)\hat{V}_{z\pi+}(-\pi + w_1)) \tag{6.7.43}$$

$$i_{d1}(w_1) = -\frac{k_1 \sin\beta_1(-Z_1 \sin w_1 \hat{I}_{z\pi+}(-\pi + w_1) + \cos(w_1)\cos\beta_1 \hat{V}_{z\pi+}(-\pi + w_1) + \sin\beta_1 \hat{V}_{\rho\pi+}(-\pi + w_1))}{2Z_1} \tag{6.7.44}$$

The problem of the diffraction of a dielectric wedge has been solved efficiently by using the Fredholm factorization, as indicated in this section. To evaluate the diffraction coefficients at skew incidence, we refer to Daniele (2009).

6.8. Criteria for examples of the scattering by a dielectric wedge at normal incidence (Ez-polarization)

In this section, we present numerical results for scattering by a dielectric wedge at normal incidence (Ez-polarization).

In order to describe the problems, we use a modified version of the criteria discussed in section 5.8:

1. aperture angle of the wedge in terms of Φ;

2. relative dielectric constant ε_r;

3. propagation constant $k = k' - jk''$ of the free space;

4. parameter for the Ez-polarized plane wave

   a. azimuthal angle of incidence φ_o

   b. intensity of plane wave Eo;

5. discretization parameters for the FIE: truncation A and step h in free space and A1, h1 for the dielectric region.

The angles are always assumed in radians.

As also stated in this section, we can select without loss of generality $k = k' - jk'' = 1 - j10^{-8}$. Spectra for different values of k are re-scaled according to a different value of k' (e.g. $k' = p$ such that $k_{new} = p(1 - j10^{-8})$, which changes the computed spectrum $F_+(\eta)$ into $(k / k_{new})F_+(k\eta / k_{new})$; see sections 2.8 and 6.5. The parameters of validation are relative errors in self-converge tests since practically no solution is available in the literature to make comparison.

The following example is taken from Daniele and Lombardi (2011) with copyright permission. For further examples, see Daniele and Lombardi (2011).

6.9. Example: the scattering by a dielectric wedge at normal incidence (Ez-polarization)

This test case analyzes all the properties of our solution in terms of spectral quantities, diffraction coefficients and total fields. According to Figure 6.1.1, the physical parameters of the problem are: $\Phi = 3\pi/4, \varepsilon_r = 3, \varphi_o = \pi/8, E_o = 1V/m$. According to GO, the Ez-polarized incident plane wave impinges on the dielectric wedge and generates two reflected waves and two transmitted waves. The two transmitted waves are not reflected in the interior region.

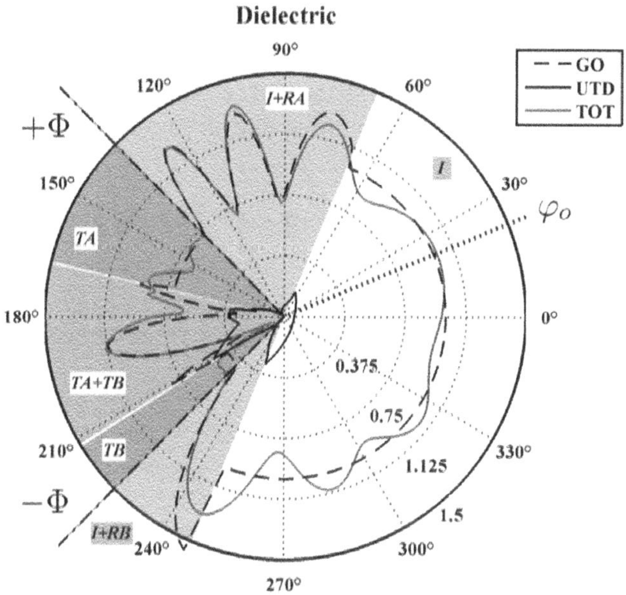

Figure 6.9.1. *GO field, UTD component and the total far-field pattern at* $k\rho = 10$ *obtained from the method proposed in this chapter with discretization parameters A = 10 and h = 0.05 (© 2011 IEEE[1]). For a higher definition version of this figure, see www.iste.co.uk/daniele/scattering2.zip*

1 Reprinted with permission from V.G. Daniele, G. Lombardi, The Wiener-Hopf solution of the isotropic penetrable wedge problem: Diffraction and total field, *IEEE Trans. Antennas Propagat.*, vol. 59, no. 10, pp. 3797–3818, 2011.

This configuration allows us to define four geometrical optics shadow boundaries: face a reflected shadow boundary, face b reflected shadow boundary, face a transmitted shadow boundary and face b transmitted shadow boundary. No incident shadow boundary exists.

In the following, we use capitalized (non-capitalized) subscripts for outgoing (incoming) azimuthal directions φ.

As shown in Figure 6.9.1 with different gray color backgrounds, there are six GO regions: incident wave region $\varphi_{RB} < \varphi < \varphi_{RA}$ (I), incident and face a reflected waves region $\varphi_{RA} < \varphi < \Phi$ (I+RA), incident and face b reflected waves region $-\Phi < \varphi < \varphi_{RB}$ (I+RB), face a transmitted wave region $\Phi < \varphi < \varphi_{TA}$ (TA), face b transmitted wave region $\varphi_{TB} < \varphi < -\Phi$ (TB) and face a and b transmitted waves region $(-\pi < \varphi < \varphi_{TB}) \bigcup (\varphi_{TA} < \varphi < \pi)$ (TA+TB).

Figure 6.9.1 presents the GO field, the UTD component and the total far field at the distance $k\rho = 10$ from the edge of the wedge. According to GO, the problem under examination shows:

– a face a reflected wave angle $\varphi_{RA} = -\pi - \varphi_o + 2\Phi = 3\pi/8 \simeq 1.18_{rad} = 67.6^o$

– a face b reflected wave angle $\varphi_{RB} = \pi - \varphi_o - 2\Phi = -5\pi/8 \simeq -1.96_{rad} = -112.3^o$

– a face a transmitted wave angle $\varphi_{TA} = -2\pi - \Phi - \arccos\left(\cos(\varphi_o + \Phi_1)/\sqrt{\varepsilon_r}\right)$ $\simeq -2.58_{rad} = -147.8^o$

– a face b transmitted wave angle $\varphi_{TB} = -2\pi - \Phi - \arccos\left(\cos(-\varphi_o + \Phi_1)\sqrt{\varepsilon_r}\right)$ $\simeq 2.92_{rad} = 167.3^o$

The solution of the problem is obtained by applying the discretization method presented in section 6.5 and in Daniele and Lombardi (2011) to the IEs (6.3.23)–(6.3.28) with representations (6.3.29)–(6.3.32) and the analytic continuations (6.4.13)–(6.4.16). It yields the axial spectra $\hat{V}_d(w)$, $\hat{V}_{\pi d}(w_1)$, $\hat{I}_+(w)$, $\hat{I}_{\pi +}(w_1)$ (we have omitted "o" in the subscripts of the outer axial spectra) of the problem that is useful for asymptotic evaluation, as discussed in section 6.6.

In this example, we apply sample and old quadrature with discretization parameters A = A1, h = h1.

Figure 6.9.2 shows the behavior of the numerical solution in terms of the spectral unknowns $\hat{V}_+(w)$ and $\hat{I}_+(w)$ in the regularity segment $-\Phi \leq w \leq 0$ that belongs to the regularity strip: the outer axial starting spectra is purely imaginary; see section 6.5 and equation (6.5.5).

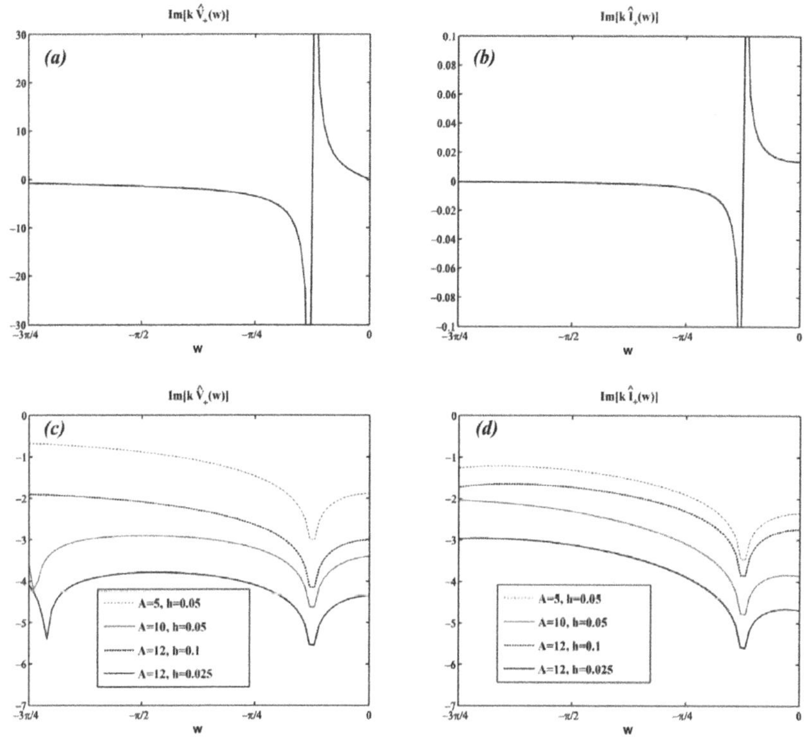

Figure 6.9.2. *a, b) Imaginary parts* $Im[k\hat{V}_+(w)]$ *and* $Im[k\hat{I}_+(w)]$ *in the regularity segment* $-\Phi \leq w \leq 0$ *that belongs to the regularity strip. c, d) Relative error expressed in the log10 scale of* $Im[k\hat{V}_+(w)]$ *and* $Im[k\hat{I}_+(w)]$ *in terms of discretization parameters A and h with respect to the reference solution obtained for A = 12, h = 0.015 (© 2011 IEEE[2]). For a higher definition version of this figure, see www.iste.co.uk/daniele/scattering2.zip*

Relative errors are expressed in log10 scale by considering as a reference solution the one obtained for discretization parameters A = 12, h = 0.015. For the

2 Reprinted with permission from V.G. Daniele, G. Lombardi, The Wiener-Hopf Solution of the Isotropic Penetrable Wedge Problem: Diffraction and Total Field, *IEEE Trans. Antennas Propagat.*, vol. 59, no. 10, pp. 3797–3818, 2011.

numerical solution, we have chosen different values of the integration parameter A and h in order to confirm the convergence of our technique. However, an excessive value of A (A>15) yields ill-conditioned matrices in the discretization process. Figure 6.9.3 shows the behavior of the numerical solution in terms of the spectral unknowns $\hat{V}_{\pi+}(w_1)$ and $\hat{I}_{\pi+}(w_1)$ in the regularity segment $-\Phi_1 \leq w_1 \leq 0$.

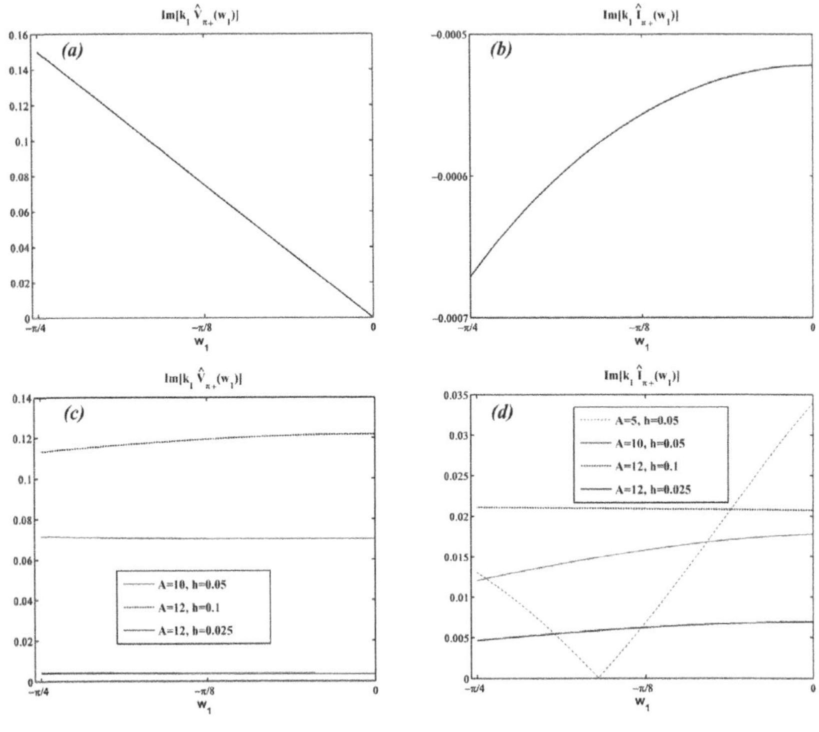

Figure 6.9.3. *a, b) Imaginary parts* $Im[k_1\hat{V}_{\pi+}(w_1)]$ *and* $Im[k_1\hat{I}_{\pi+}(w_1)]$ *in the regularity segment* $-\Phi_1 \leq w_1 \leq 0$ *that belongs to the regularity strip. c, d) Relative error expressed in the log10 scale of* $Im[k_1\hat{V}_{\pi+}(w_1)]$ *and* $Im[k_1\hat{I}_{\pi+}(w_1)]$ *in terms of discretization parameters A and h with respect to the reference solution obtained for A = 12, h = 0.015 (© 2011 IEEE[3]). For a higher definition version of this figure, see www.iste.co.uk/daniele/scattering2.zip*

3 Reprinted with permission from V.G. Daniele, G. Lombardi, The Wiener-Hopf Solution of the Isotropic Penetrable Wedge Problem: Diffraction and Total Field, *IEEE Trans. Antennas Propagat.*, vol. 59, no. 10, pp. 3797–3818, 2011.

As discussed in section 6.6, the evaluation of the GTD coefficients requires the analytic continuation of the spectral unknowns $\{\hat{V}_+(w), \hat{I}_+(w)\}$ and $\{\hat{V}_{\pi+}(w_1), \hat{I}_{\pi+}(w_1)\}$ respectively in the interval $-\pi - \Phi < w < -\pi + \Phi$ and $-\pi - \Phi_1 < w_1 < -\pi + \Phi_1$. The required analytic continuation is obtained through the recursive equations presented in section 6.4.

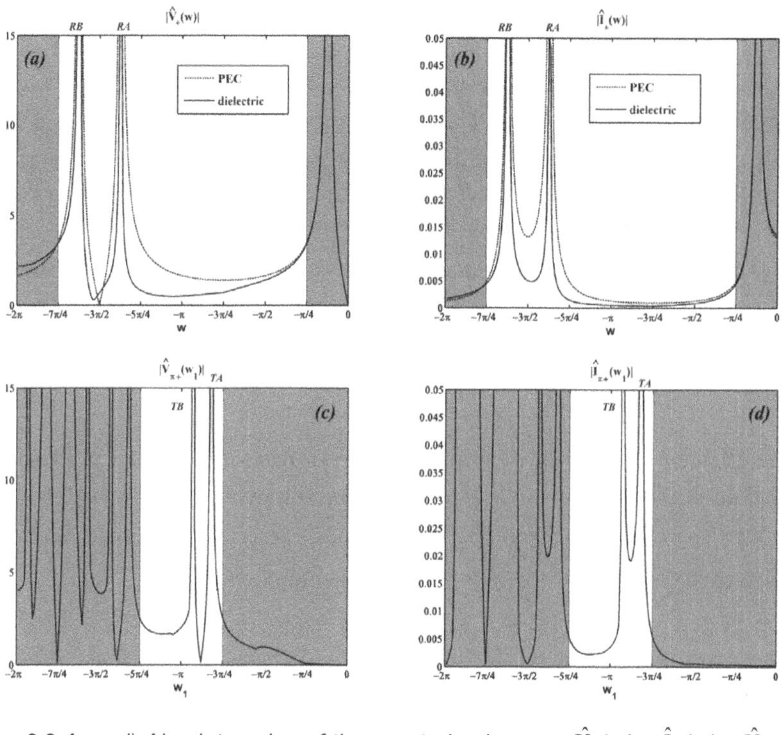

Figure 6.9.4. *a, d) Absolute value of the spectral unknowns $\hat{V}_+(w)$, $\hat{I}_+(w)$, $\hat{V}_{\pi+}(w_1)$ and $\hat{I}_{\pi+}(w_1)$ in $(-2\pi, 0)$ after the recursions. The dark gray regions are not used to evaluate the GTD coefficients. Discretization parameters A = 10 and h = 0.05 (© 2011 IEEE[4]). For a higher definition version of this figure, see www.iste.co.uk/daniele/scattering2.zip*

Figure 6.9.4 shows the behavior of the absolute value of the spectral unknowns $\hat{V}_+(w)$, $\hat{I}_+(w)$ and $\hat{V}_{\pi+}(w_1)$, $\hat{I}_{\pi+}(w_1)$ after the recursions. The figure highlights the

4 Reprinted with permission from V.G. Daniele, G. Lombardi, The Wiener-Hopf Solution of the Isotropic Penetrable Wedge Problem: Diffraction and Total Field, *IEEE Trans. Antennas Propagat.*, vol. 59, no. 10, pp. 3797–3818, 2011.

spectral regions necessary to evaluate the GTD diffraction coefficients outside and inside the wedge, i.e. $-\pi-\Phi < w < -\pi+\Phi$ and $-\pi-\Phi_1 < w_1 < -\pi+\Phi_1$. The figure also shows the GO poles relevant to the GTD.

In particular, $\{\hat{V}_+(w), \hat{I}_+(w)\}$ show peaks for the reflected waves, while $\{\hat{V}_{\pi+}(w_1), \hat{I}_{\pi+}(w_1)\}$ for the transmitted waves. The location of the poles agrees with the standard GO theory, for example, consider $w_{RA} \simeq -4.32$ and $w_{TB} \simeq -2.92$: using the even property of plus functions and the recursive equations, from the GTD expressions (section 6.6) in terms of axial spectral unknowns/Sommerfeld functions, we obtain $\varphi_{RA} = -w_{RA} - \pi \simeq 1.18 rad$ and $\varphi_{TB} = -w_{TB} \simeq 2.92 rad$ (since $\varphi_1 = \pi - \varphi$). Similar considerations hold for the other GO poles. The GO components can be obtained by using standard GO techniques or by applying the SDP method (sections 6.6 and 5.7). Note that the study of the axial spectra is fundamental. In fact, if w_{oi} is a singularity of $\hat{V}_+(w)$ ($\hat{I}_+(w)$), the spectrum of $\hat{V}_+(w,\varphi)$ ($\hat{I}_+(w,\varphi)$) presents the singularities $w_{oi} \pm \varphi$ that can be captured by the integration contour deformation from the Bromwich contour B_r to SDP, while applying SDP method (sections 6.6 and 5.7).

Investigating the spectral properties of the functions, from their numerical evaluation, while $k\hat{V}_+(w)$ and $k\hat{I}_+(w)$ are almost purely imaginary functions in $(-2\pi, 0)$ (purely imaginary in $(-\Phi, 0)$, as described in section 6.5 and equation (6.5.5)), $k_1\hat{V}_{\pi+}(w_1)$ and $k_1\hat{I}_{\pi+}(w_1)$ are complex functions.

The approximate total GTD diffraction coefficients are estimated substituting the approximations of the spectral unknowns (sections 6.6 and 5.7). Figure 6.9.5.a presents the absolute value of the total GTD diffraction coefficient (in dB) for each observation angle φ. The peaks of the GTD diffraction coefficients occur for the GO angles: reflected and transmitted waves. The convergence is shown in Figure 6.9.5.b for different integration parameters through the evaluation of the relative error in log10 scale with respect to the reference solution obtained for A = 12, h = 0.015. The log10 scale measures the level of precision in terms of digits for each observation angle φ. Figure 6.9.5.c shows the phase of the total GTD diffraction coefficient (in dB) for each observation angle φ. Figure 6.9.5a and c also shows the plots for the PEC wedge for comparison.

The complete solution is shown in Figure 6.9.6. On the left, we present the total field, GO field component, UTD field component at the distance $k\rho = 10$. UTD is computed from GTD, as discussed in section 5.7. The gray regions are inside the wedge. On the right, we show the comparison between the total field of the dielectric wedge with the one of the PEC wedge. We note that Figure 6.9.6 (right) shows a small loss of convergence for $\varphi \simeq +2.97$ and $\varphi \simeq -2.36$: the reasons for the corner behavior are different. In the first case, when $\varphi \simeq +2.97$, the problem is due to the spectral reconstruction of the Wiener-Hopf inner axial spectral unknowns (π unknowns) in w_1, i.e. $\hat{V}_{\pi+}(w_1)$ and $\hat{I}_{\pi+}(w_1)$. In fact, for $\varphi \simeq +2.97$, the UTD field component is related to the evaluation of $\hat{V}_{\pi+}(w_1)$ and $\hat{I}_{\pi+}(w_1)$ in $w_1 = -2.97$; see GTD diffraction coefficient (6.6.10). We recall that the recursive equations are used to estimate the spectral unknowns out of the regularity strips ($-\Phi \leq \text{Re}[w] \leq 0$ and $-\Phi_1 \leq \text{Re}[w_1] \leq 0$), in particular in the GTD intervals ($-\pi - \Phi < w < -\pi + \Phi$ and $-\pi - \Phi_1 < w_1 < -\pi + \Phi_1$). Figure 6.9.8 shows the mapping used in the estimation of π functions out of the regularity segment $-\Phi_1 \leq w_1 \leq 0$ for the real value of w_1 (see 6.4.13)–(6.4.16)), i.e. $w = \Phi + g_1(w_1 + \Phi_1)$. The map starts from $w_1 = -\Phi_1$ that yields $w = \Phi - \arccos(\sqrt{\varepsilon_r}) \simeq 3\pi/4 - 1.1462$ and goes all over the gray line as long as $w_1 = -\pi - \Phi_1$ is mapped into $w = -\Phi_1 + \arccos\ (\sqrt{\varepsilon_r}) \simeq -\pi/4 + j1.1462$. Uniform sampling for real w_1 is mapped into non-uniform sampling in complex w. In particular, the significant point $w_1 = -2.97$ is mapped into $w = -\Phi_1$, where the mapping shows a change in slope and highly non-uniform sampling. This is the cause for the loss of convergence in UTD at $\varphi \simeq +2.97$.

The second direction where we experience loss of convergence is $\varphi \simeq -2.36$ (Figure 6.9.6, left), and it is due to the UTD uniform expression of the GTD field component. Since $\varphi \simeq -2.36$ is very close to the interface between the two materials ($\varphi = -\Phi$), the Kouyoumjian–Pathak (KP) transition function (see section 5.7 for definition) is not adequate to model the problem: the uniform diffraction component is a cylindrical wave whose intensity should vanish at the interface; on the contrary, at first sight, the slope of the KP transition function does not consider the change of materials.

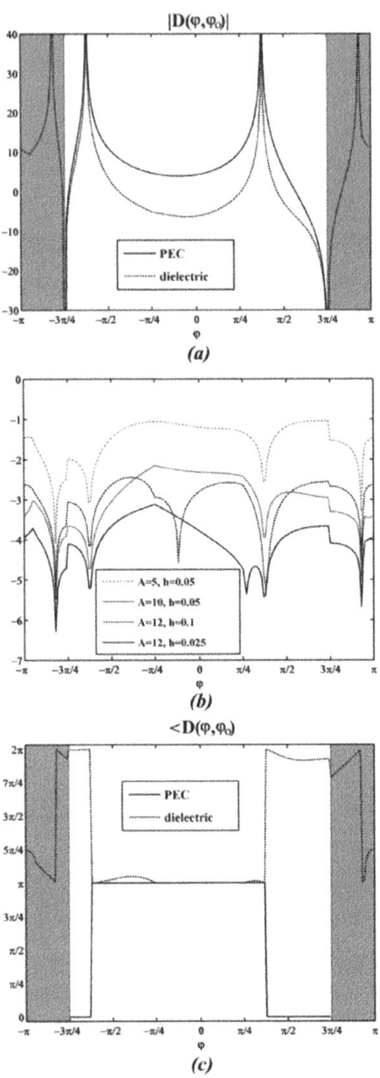

Figure 6.9.5. *a) Absolute value of the total GTD diffraction coefficient (dB), b) GTD diffraction's relative error in log10 scale using different sets of integration parameters (reference solution A = 12, h = 0.015) and c) phase of the total GTD diffraction coefficient. Subfigure a and c also report the solution of PEC wedge for comparison (© 2011 IEEE[5]). For a higher definition version of this figure, see www.iste.co.uk/daniele/scattering2.zip*

---

5 Reprinted with permission from V.G. Daniele, G. Lombardi, The Wiener-Hopf Solution of the Isotropic Penetrable Wedge Problem: Diffraction and Total Field, *IEEE Trans. Antennas Propagat.*, vol. 59, no. 10, pp. 3797–3818, 2011.

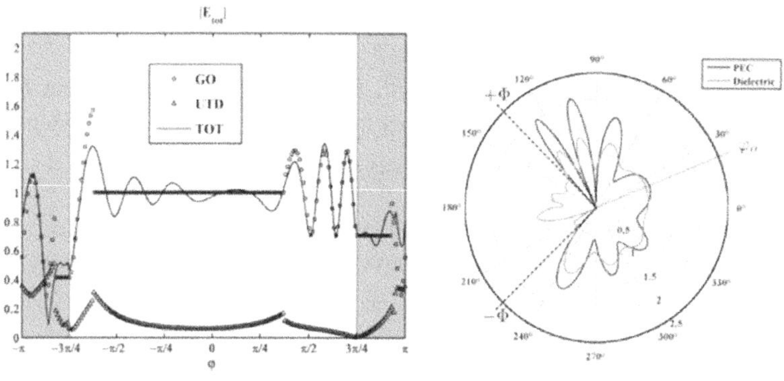

Figure 6.9.6. *Left: total field (solid line), GO field component (squares), UTD field component (triangles) at $k\rho = 10$. Right: comparison between the total field of the dielectric (gray line) and the PEC wedge (black line) at $k\rho = 10$ (figure modified and adapted from Daniele and Lombardi (2011)). For a higher definition version of this figure, see www.iste.co.uk/daniele/scattering2.zip*

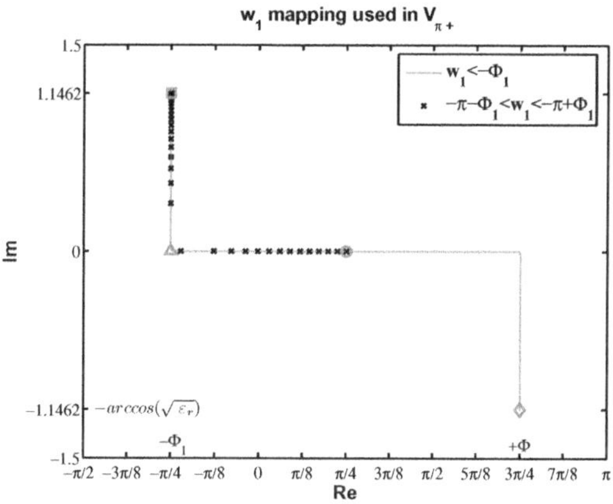

Figure 6.9.7. $w = \Phi + g_1(w_1 + \Phi_1)$ *mapping used for the analytical continuation of* π *functions out of the regularity segment* $-\Phi_1 \le w_1 \le 0$. *The map starts from* $w_1 = -\Phi_1$ *that yields* $w = \Phi - \arccos(\sqrt{\varepsilon_r})$ ($\arccos(\sqrt{\varepsilon_r}) \simeq j1.1462$) *and goes all over the gray line as long as* $w_1 = -\pi - \Phi_1$ *is mapped into* $w = -\Phi_1 + \arccos(\sqrt{\varepsilon_r})$. *Uniform sampling for real* w_1 *corresponds to non-uniform sampling in complex* w. *The symbols* ◇, ,△,□ *respectively are the mapped values of* $w_1 = -\Phi_1, \Phi_1 - \pi, -\Phi_1 - \arccos(-1/\sqrt{\varepsilon_r}), -\pi - \Phi_1$ *(© 2011 IEEE[6]). For a higher definition version of this figure, see www.iste.co.uk/daniele/scattering2.zip*

6 Reprinted with permission from V.G. Daniele, G. Lombardi, The Wiener-Hopf Solution of the Isotropic Penetrable Wedge Problem: Diffraction and Total Field, *IEEE Trans. Antennas Propagat.*, vol. 59, no. 10, pp. 3797–3818, 2011.

6.10. Discussion

The GWHE formulation together with the direct application of the Fredholm factorization method in $\bar{w}, \bar{w}_1$ constitutes a tool to solve general arbitrary penetrable wedge problems illuminated by plane waves with any aperture angle.

In this chapter, we have analyzed in depth the electromagnetic dielectric wedge problem at normal incidence and we have provided the general theory at skew incidence.

In particular, we have shown all the capability of the method with an in-depth analysis on analytic properties of functions and we have shown the performance of the spectral approximations. A practical problem of scattering by penetrable wedge is presented for physical and engineering applications. The example is studied in detail to make the reader capable of independent implementation using technical computing software, in particular in analytic and numerical manipulations of the equations. The proposed method can be extended to other physics such as elasticity and acoustics as already demonstrated in simple problems in the previous chapters.

Appendix 6.A. Fredholm factorization applied to (6.3.2)–(6.3.5)

Taking inspiration from Daniele (2009, 2010) and Daniele and Lombardi (2011), starting from the GWHEs (6.3.2)–(6.3.5) in the normalized form of the problem, we apply the mappings $\bar{\eta}$ and $\bar{\eta}_d$ (6.1.4)–(6.1.5) yielding

$$\bar{Y}_{1+}(\bar{\eta}) = \bar{X}_{1-}(\bar{\eta}) - \frac{\bar{\xi}_-}{\bar{n}_+}\bar{X}_{2-}(\bar{\eta}) \tag{6.A.1}$$

$$\ddot{Y}_{2+}(\bar{\eta}_d) = \ddot{X}_{1-}(\bar{\eta}_d) + \frac{\ddot{\xi}_{1-}}{\ddot{n}_{1+}}\ddot{X}_{2-}(\bar{\eta}_d) \tag{6.A.2}$$

$$\bar{Y}_{3+}(\bar{\eta}) = \bar{X}_{3-}(-\bar{\eta}) - \frac{\bar{\xi}_-}{\bar{n}_+}\bar{X}_{4-}(-\bar{\eta}) \tag{6.A.3}$$

$$\ddot{Y}_{4+}(\bar{\eta}_d) = \ddot{X}_{3-}(\bar{\eta}_d) + \frac{\ddot{\xi}_{1-}}{\ddot{n}_{1+}}\ddot{X}_{4-}(\bar{\eta}_d) \tag{6.A.4}$$

where $\bar{Y}_{i+}(\bar{\eta}) = Y_{i+}(\eta), \bar{X}_{i-}(\bar{\eta}) = X_{i+}(-m), \ddot{Y}_{2+}(\bar{\eta}_d) = \dot{Y}_{i+}(\eta), \ddot{X}_{1-}(\bar{\eta}_d) = \dot{X}_{i+}(-m_1)$ and see also section 6.3 for definitions. Note that the WH equations (6.A.1)–(6.A.4) are defined into two different complex planes $\bar{\eta}$ and $\bar{\eta}_d$ in normal form.

The application of the Fredholm factorization method to (6.A.1)–(6.A.4) allows us to eliminate the plus functions

$$\begin{cases} \overline{X}_{(i)-}(\overline{\eta})+\dfrac{k_o+\overline{\eta}}{\sqrt{k_o^2-\overline{\eta}^2}}\overline{X}_{(i+1)-}(\overline{\eta})-\dfrac{1}{2\pi j}\displaystyle\int_{-\infty}^{\infty}\dfrac{\left(\dfrac{k_o+\overline{\eta}'}{\sqrt{k_o^2-\overline{\eta}'^2}}-\dfrac{k_o+\overline{\eta}}{\sqrt{k_o^2-\overline{\eta}^2}}\right)\overline{X}_{(i+1)-}(\overline{\eta}')}{\overline{\eta}'-\overline{\eta}}d\overline{\eta}'=\overline{n}_i(\overline{\eta}) \\ \ddot{X}_{(i)-}(\overline{\eta}_d)-\dfrac{k_1+\overline{\eta}_d}{\sqrt{k_1^2-\overline{\eta}_d^2}}\ddot{X}_{(i+1)-}(\overline{\eta}_d)+\dfrac{1}{2\pi j}\displaystyle\int_{-\infty}^{\infty}\dfrac{\left(\dfrac{k_1+\overline{\eta}_d'}{\sqrt{k_1^2-\overline{\eta}_d'^2}}-\dfrac{k_1+\overline{\eta}_d}{\sqrt{k_1^2-\overline{\eta}_d^2}}\right)\dot{X}_{(i+1)-}(\overline{\eta}_d')}{\overline{\eta}_d'-\overline{\eta}_d}d\overline{\eta}_d'=0 \end{cases}$$

(6.A.5)

with $i=1,3$ and where $\overline{n}_i(\overline{\eta})$ is related to the non-standard component of the unknown functions in (6.A.1)–(6.A.4) (see Appendix 6.B):

$$\overline{n}_i(\overline{\eta})=\begin{cases} \overline{Y}_{i+}^{n.s.}(\overline{\eta}) & 0<\varphi_o<\Phi/2 \\ \overline{X}_{(i)-}^{n.s.}(\overline{\eta})+\dfrac{k_o+\overline{\eta}}{\sqrt{k_o^2-\overline{\eta}^2}}\overline{X}_{(i+1)-}^{n.s.}(\overline{\eta}) & \Phi/2<\varphi_o<\Phi \end{cases} \quad (6.A.6)$$

with φ_o the incoming direction of the source constituted of an incident plane wave.

From (6.A.5) and (6.A.1)–(6.A.4), we also obtain the representations of plus unknowns

$$\begin{cases} \overline{Y}_{i+}(\overline{\eta})=\overline{X}_{(i)-}(\overline{\eta})+\dfrac{k_o+\overline{\eta}}{\sqrt{k_o^2-\overline{\eta}^2}}\overline{X}_{(i+1)-}(\overline{\eta})=\dfrac{1}{2\pi j}\displaystyle\int_{-\infty}^{\infty}\dfrac{\left(\dfrac{k_o+\overline{\eta}'}{\sqrt{k_o^2-\overline{\eta}'^2}}-\dfrac{k_o+\overline{\eta}}{\sqrt{k_o^2-\overline{\eta}^2}}\right)\overline{X}_{(i+1)-}(\overline{\eta}')}{\overline{\eta}'-\overline{\eta}}d\overline{\eta}'+\overline{n}_i(\overline{\eta}) \\ \ddot{Y}_{(i+1)+}(\overline{\eta}_d)=\ddot{X}_{(i)-}(\overline{\eta}_d)-\dfrac{k_1+\overline{\eta}_d}{\sqrt{k_1^2-\overline{\eta}_d^2}}\ddot{X}_{(i+1)-}(\overline{\eta}_d)=-\dfrac{1}{2\pi j}\displaystyle\int_{-\infty}^{\infty}\dfrac{\left(\dfrac{k_1+\overline{\eta}_d'}{\sqrt{k_1^2-\overline{\eta}_d'^2}}-\dfrac{k_1+\overline{\eta}_d}{\sqrt{k_1^2-\overline{\eta}_d^2}}\right)\dot{X}_{(i+1)-}(\overline{\eta}_d')}{\overline{\eta}_d'-\overline{\eta}_d}d\overline{\eta}_d' \end{cases}$$

(6.A.7)

with $i=1,3$.

As considered in Chapters 1 and 5, it is convenient to introduce the complex planes $\bar{w}$ and $\bar{w}_d$ defined by

$$\bar{\eta} = -k_o \cos \bar{w}, \quad \bar{\eta}_d = -k_d \cos \bar{w}_d \tag{6.A.8}$$

In particular, we apply the warping of the integration lines on the straight lines $u \in \mathbb{R}$ and $v \in \mathbb{R}$ defined by:

$$\bar{w} = -\frac{\pi}{2} + ju, \quad \bar{w}_d = -\frac{\pi}{2} + jv \tag{6.A.9}$$

Finally, we obtain from (6.A.5)

$$\begin{cases} P_i(u) - \tan\dfrac{\dfrac{-\pi}{2} + ju}{2} P_{i+1}(u) - \dfrac{1}{\pi j}\displaystyle\int_{-\infty}^{\infty} M(u,u') P_{i+1}(u')\, du' = n_i(u) \\ Q_i(v) + \tan\dfrac{-\pi^2 + jv}{2} Q_{i+1}(v) + \dfrac{1}{\pi j}\displaystyle\int_{-\infty}^{\infty} M(v,v') Q_{i+1}(v')\, dv' = 0 \end{cases} \tag{6.A.10}$$

for $u \in R, v \in R$, $i = 1,3$ and where

$$\begin{cases} P_i(u) = k_o X_{i+}\left(k_o \cos\left[\dfrac{\Phi}{\pi}\left(-\dfrac{\pi}{2} + ju \right) + \Phi \right] \right) \\ Q_i(v) = k_o \dot{X}_{i+}\left(k_1 \cos\left[\dfrac{\Phi_1}{\pi}\left(-\dfrac{\pi}{2} + jv \right) + \Phi_1 \right] \right) \\ M(u,u') = \dfrac{e^u}{(e^u - j)} \dfrac{e^{u'} + j}{e^{u+u'} + 1} \\ n_i(u) = \bar{n}_i(\bar{\eta}(u)) \end{cases} \tag{6.A.11}$$

for $i = 1,2,3,4$.

Appendix 6.B. Source term $\bar{n}_i(\bar{\eta})$

In this Appendix, we evaluate the source term $\bar{n}_i(\bar{\eta})$ (6.A.6) of the FIEs (6.A.5) in $\bar{\eta}$ and thus in u (6.A.10), as discussed in section 6.3 in (6.3.8) and (6.3.10).

With reference to Figure 6.1.1, we assume the propagation constant k_o with small vanishing losses, and for simplicity, we consider initially an obtuse aperture angle $\Phi > \pi/2$.

The physical source is constituted of an Ez-polarized incident plane wave with azimuthal direction of incidence $0 < \varphi_o < \Phi$ (6.B.1) that yields a first-order pole $\bar{\eta}_o = -k_o \cos((\pi/\Phi)\varphi_o)$ in the axial spectral unknowns of the free space region:

$$E_z^i = E_o u(\pi - |\varphi - \varphi_o|) e^{jk_o \rho \cos(\varphi - \varphi_o)} \tag{6.B.1}$$

The GO field outside the wedge is

$$\begin{cases} E_z^g = E_o u(\pi - |\varphi - \varphi_o|) e^{jk_o \rho \cos(\varphi - \varphi_o)} + u(\pi - |\varphi + \varphi_o - 2\Phi|) E_{or}^{(a)} e^{jk_o \rho \cos(\varphi + \varphi_o - 2\Phi)} \\ \quad + u(\pi - |\varphi + \varphi_o + 2\Phi|) E_{or}^{(b)} e^{jk_o \rho \cos(\varphi + \varphi_o + 2\Phi)} + E_{zt}^g \\ H_\rho^g = j \dfrac{1}{k_o Z_o \rho} \dfrac{\partial E_z^g}{\partial \varphi} \end{cases} \tag{6.B.2}$$

where $E_{or}^{(a,b)}$ are the intensities of the waves reflected by the faces a and b with incoming direction respectively $\varphi_{ra} = 2\Phi - \varphi_o$ and $\varphi_{rb} = 2\Phi + \varphi_o$, and E_{zt}^g is the field re-transmitted from inside the wedge when it is present with incoming directions φ_{tn} (see Kim *et al.* (1991a) for an in-depth discussion).

In order to establish the source term $\bar{n}_i(\bar{\eta})$, we need to collect the non-standard part of plus and minus unknowns in the $\bar{\eta}$ plane associated with poles located in the principal proper sheet, i.e. in $-\Phi < w < 0$.

We also recall that a plus (minus) function in the $\bar{\eta}$ plane is non standard with poles in $-\Phi/2 < w < 0$ ($-\Phi < w < -\Phi/2$) and its non-standard part corresponds to the non-standard part of its GO component.

Since (6.A.1)–(6.A.4) are written in different complex planes $\bar{\eta}$ and $\bar{\eta}_d$, we cannot use the classical property demonstrated for impenetrable wedge problems, where, once the GO component of plus function in the proper sheet (i.e. just the incident field) is known, we obtain the GO component of minus functions applying the CWHE to the GO component of plus functions (see section 5.3).

For this reason, we need to investigate the regularity properties of the plus and minus spectra of the GO field (this is a general fundamental way to determine the non-standard parts).

In particular, we need to evaluate the Laplace transforms of the GO field along $\varphi = 0, \Phi, -\Phi$ respectively in domain η, m, m that with reference to Figure 3.3.1 are defined as $F_+(\eta, \varphi = 0) = \int_0^\infty f_+(\rho, \varphi = 0) e^{+j\eta\rho} d\rho$ with $\rho = x$ and $F_+(-m, \varphi = \pm\Phi) = \int_0^\infty f_-(\rho, \varphi = \pm\Phi) e^{-jm\rho} d\rho$ with $\rho = v$.

The poles need to be analyzed in $\bar{\eta}$ and w planes to understand their non-standard properties and the sheet where they are located.

Let us start from the plus axial spectra $\varphi = 0$.

We recall that $\varphi_o, \varphi_{ra}, \varphi_{rb}$ yields different poles in the η plane ($\eta_{o,ra,rb} = -k_o \cos(\varphi_{o,ra,rb})$) mapped into the same point $\bar{\eta}_o = -k_o \cos((\pi/\Phi)\varphi_o)$ in the $\bar{\eta}$ plane but in different sheets, as the corresponding points in the w plane are $w_{o,ra,rb} = -\varphi_{o,ra,rb}$. Re-transmitted waves generate further poles $\bar{\eta}_{tn} = -k_o \cos((\pi/\Phi)\varphi_{tn})$ that correspond to w_{tn} after GO consideration in the w plane.

In this case, as stated in section 5.3, the incident pole lies in the principal proper sheet while the reflected waves together with the re-transmitted poles are not in this sheet ($w < -\Phi$). The incident wave is a non-standard component of the axial spectra only if $0 < \varphi_o < \Phi/2$ as $\mathrm{Im}[\bar{\eta}_o] > 0$.

We also note that we have the same non-standard spectral properties of plus unknowns for incident waves with incoming directions $0 < \varphi_o < \Phi$ and $-\Phi < \varphi_o < 0$, i.e. same poles in spectral planes $\bar{\eta}$ and w, as the two types of directions are symmetric with respect to positive x axis, which is the reference for the definition of the Laplace transform in η ($\varphi = 0$).

In summary, the non-standard part to be considered for $0 < \varphi_o < \Phi/2$ is the one of the plus unknowns and it is constituted only of the incident field.

Now we investigate the more cumbersome case of non-standard minus functions.

In this case, we need to consider the minus functions defined as m-Laplace transforms of field component (6.B.3) along the faces of the wedge $\varphi = \Phi, -\Phi$.

Let us begin from the analysis of the spectra for $\varphi = \Phi$.

It yields for the incident, face a reflected and face b reflected, waves the poles in the m plane:

$$\begin{aligned} m_o^{(+\Phi)} &= k_o \cos(\Phi - \varphi_o) \\ m_{ra}^{(+\Phi)} &= k_o \cos(\Phi + \varphi_o - 2\Phi) = k_o \cos(\Phi - \varphi_o) \\ m_{rb}^{(+\Phi)} &= k_o \cos(\Phi + \varphi_o + 2\Phi) = k_o \cos(3\Phi + \varphi_o) \end{aligned} \tag{6.B.3}$$

Moreover, in this case, the poles (6.B.3) are mapped into the same point $\bar{\eta}_o = -k_o \cos((\pi / \Phi)\varphi_o)$ in the $\bar{\eta}$ plane.

Further poles are generated by re-transmitted waves $m_{tn}^{(+\Phi)} = k_o \cos(\Phi - \varphi_{tn})$.

We note that, because of the symmetry with respect to the direction $\varphi = \Phi$, the incident and face a reflected waves generate the same pole in the same sheet of Laplace domain m, i.e. their poles are coincident. From

$$m = m(\eta) = -\eta \cos\Phi + \xi(\eta) \sin\Phi = k_o \cos(w + \Phi) \tag{6.B.4}$$

in the w plane (6.B.3) corresponds to

$$\begin{aligned} w_o^{(+\Phi)} &= -\varphi_o \\ w_{ra}^{(+\Phi)} &= -\varphi_o \\ w_{rb}^{(+\Phi)} &= +\varphi_o + 2\Phi \end{aligned} \tag{6.B.5}$$

and $w_{tn}^{(+\Phi)}$. GO considerations in the w plane yield that face b reflected pole and the re-transmitted poles are not in the principal proper sheet ($w < -\Phi$).

The incident and face a reflected waves are a non-standard component of the minus spectra only if $\Phi / 2 < \varphi_o < \Phi$ as $\mathrm{Im}[\bar{\eta}_o] < 0$.

With the same procedure, we analyze the spectra for $\varphi = -\Phi$, which yields

$$\begin{aligned} m_o^{(-\Phi)} &= k_o \cos(-\Phi - \varphi_o) = k_o \cos(\Phi + \varphi_o) \\ m_{ra}^{(-\Phi)} &= k_o \cos(-\Phi + \varphi_o - 2\Phi) = k_o \cos(3\Phi - \varphi_o) \\ m_{rb}^{(-\Phi)} &= k_o \cos(-\Phi + \varphi_o + 2\Phi) = k_o \cos(\Phi + \varphi_o) \end{aligned} \tag{6.B.6}$$

with corresponding

$$\begin{aligned} w_o^{(-\Phi)} &= +\varphi_o \\ w_{ra}^{(-\Phi)} &= 2\Phi - \varphi_o \\ w_{rb}^{(-\Phi)} &= +\varphi_o \end{aligned} \tag{6.B.7}$$

that lie in the proper sheet if $-\Phi < w < 0$ and generate a non-standard minus component if $-\Phi < w < -\Phi/2$ (it is possible only for $-\Phi < \varphi_o < 0$).

In summary, the non-standard part to be considered for $\Phi/2 < \varphi_o < \Phi$ is the one of the minus unknowns and it is constituted only of the incident and face a reflected waves.

Finally, a practical rule to discover non-standard parts is summarized here.

Assuming $0 < \varphi_o < \Phi$:

1) solve the GO problem and verify the incoming directions φ_{go};

2) for each GO wave, verify the existence (not in shadow) along the direction of Laplace transforms $\varphi = 0, \Phi, -\Phi$; in case of absence, we can ignore it;

3) let us consider only the region outside the wedge for $0 < \varphi_{go} < \Phi$ (the rule can be extended for symmetry to $-\Phi < \varphi_{go} < 0$ and modified for the region inside the wedge). If $\Phi/2 < \varphi_{go} < \Phi$ ($0 < \varphi_{go} < \Phi/2$), then the wave is a potential non-standard minus (plus) function to be considered

4) to understand if φ_{go} creates a non-standard part in the principal proper sheet that yields a contribution in $\bar{n}_i(\bar{\eta})$, verify whether the corresponding pole in the w plane satisfies $-\Phi < w_{go} < 0$.

In particular, the location of w_{go} is critical in the presence of multiple interactions (reflections and transmissions) for small Φ with $\Phi < \pi/2$. We recall that each interaction as in multiple reflected waves generates a shift of the corresponding pole in the w plane, i.e. we have $w_{go} = -\varphi_n$, where φ_n is the incoming direction defined in section 5.3.

Now we are ready to compute (6.A.6)

$$\bar{n}_i(\bar{\eta}) = \begin{cases} \bar{Y}_{i+}^{n.s.}(\bar{\eta}) & 0 < \varphi_o < \Phi/2 \\ \bar{X}_{(i)-}^{n.s.}(\bar{\eta}) + \dfrac{k_o + \bar{\eta}}{\sqrt{k_o^2 - \bar{\eta}^2}} \bar{X}_{(i+1)-}^{n.s.}(\bar{\eta}) & \Phi/2 < \varphi_o < \Phi \end{cases} \tag{6.B.8}$$

with i = 1,3.

In order to obtain the explicit form of (6.B.8), we use computations in the w plane and therefore we give some useful expressions:

$$\begin{aligned} & m = k_o \cos(w+\Phi), \quad k_1 = \tau_1 \cos(w_1 + \Phi_1) \\ & \xi = -k_o \sin w, \quad \xi_1 = -k_1 \sin w_1 \\ & \xi_- = \sqrt{\tau_o} \sin\frac{\overline{w}}{2}, \quad \xi_{1-} = \sqrt{\tau_1} \sin\frac{\overline{w}_1}{2} \\ & n = k_o \sin(w+\Phi), \quad n_1 = k_1 \sin(w_1 + \Phi_1) \\ & n_+ = \sqrt{k_o} \cos\frac{\overline{w}}{2}, \quad n_{1+} = \sqrt{k_1} \cos\frac{\overline{w}_1}{2} \end{aligned} \tag{6.B.9}$$

Let us start form the case $0 < \varphi_o < \Phi/2$. For this case, we have only the incident wave that generates non-standard part in plus functions; therefore, we need to consider the non-standard part of $\bar{Y}_{1+}(\bar{\eta}), \bar{Y}_{3+}(\bar{\eta})$

$$\bar{Y}_{i+}(\bar{\eta}) = \bar{Y}_{i+}^{s}(\bar{\eta}) + \frac{\bar{T}_{oi}}{\bar{\eta} - \bar{\eta}_o} \tag{6.B.10}$$

whose definition in the η plane are given in section 6.3

$$Y_{1+}(\eta) = 2\frac{\xi_+}{n_{o+}} V_{o+}(\eta), \quad Y_{3+}(\eta) = \frac{2\omega\mu_o}{n_{o+}\sqrt{k_o}} I_{o+}(\eta) \tag{6.B.11}$$

We recall that

$$\overline{T}_{oi} = Res[\overline{Y}_{i+}(\overline{\eta})]\Big|_{\overline{\eta}_o} = T_{oi}\left.\frac{d\overline{\eta}}{d\eta}\right|_{\eta_o} \tag{6.B.12}$$

with $T_{oi} = Res[Y_{i+}(\eta)]\Big|_{\eta_o}$ and $\eta_o = -k_o \cos\varphi_o = -k_o \cos w_o$ (incident wave pole in the η plane).

From (5.3.2) and (5.3.3)

$$V_{z+}^i(\eta,0) = \int_0^\infty E_o e^{jk_o\rho\cos(0-\varphi_o)} e^{+j\eta\rho} d\rho = \frac{jE_o}{\eta-\eta_o},$$

$$I_{\rho+}^i(\eta,0) = \int_0^\infty \frac{\sin(-\varphi_o)E_o}{Z_o} e^{jk_o\rho\cos(0-\varphi_o)} e^{-jm\rho} d\rho = \frac{\sin(-\varphi_o)}{Z_o}\frac{jE_o}{\eta-\eta_o} \tag{6.B.13}$$

and using the definitions of the involved function in the w plane (6.B.9), we obtain for (6.B.10) in detail

$$\overline{T}_{o1} = Res[\overline{Y}_{1+}(\overline{\eta})]\Big|_{\overline{\eta}_o}$$

$$= T_{o1}\left.\frac{d\overline{\eta}}{d\eta}\right|_{\eta_o} = 2\frac{\sqrt{k_o^2-\eta^2}}{-\sqrt{\frac{k_o+\overline{\eta}}{2}}}\frac{1}{\sqrt{\frac{k_o-\overline{\eta}}{2}}} jE_o \left.\frac{k_o\frac{\pi}{\Phi}\sin(\frac{\pi}{\Phi}w)}{k_o\sin(w)}\right|_{\eta_o,w_o,\overline{\eta}_o}$$

$$= 4\frac{-k_o\sin(w)}{k_o\sin(\frac{\pi}{\Phi}w)} jE_o \frac{k_o\frac{\pi}{\Phi}\sin(\frac{\pi}{\Phi}w)}{k_o\sin(w)} = -4jE_o\frac{\pi}{\Phi}$$

$$\overline{T}_{o3} = Res[\overline{Y}_{3+}(\overline{\eta})]\Big|_{\overline{\eta}_o} = T_{o3}\left.\frac{d\overline{\eta}}{d\eta}\right|_{\eta_o}$$

$$= 2\frac{\omega\mu_o}{\sqrt{\frac{k_o-\overline{\eta}}{2}}\sqrt{k_o}}(-j\sin\varphi_o E_o/Z_o)\left.\frac{k_o\frac{\pi}{\Phi}\sin(\frac{\pi}{\Phi}w)}{k_o\sin(w)}\right|_{\eta_o,w_o,\overline{\eta}_o} \tag{6.B.14}$$

$$= 2\frac{-jk_o\sin\varphi_o E_o}{k_o\cos(\frac{\pi}{2\Phi}w)}\left.\frac{k_o\frac{\pi}{\Phi}\sin(\frac{\pi}{\Phi}w)}{k_o\sin(w)}\right|_{\eta_o,w_o,\overline{\eta}_o} = -4jE_o\frac{\pi}{\Phi}\sin(\frac{\pi\varphi_o}{2\Phi})$$

where $\dfrac{d\bar{\eta}}{d\eta} = \dfrac{d\bar{\eta}}{dw}\dfrac{dw}{d\eta} = \dfrac{k_o \dfrac{\pi}{\Phi}\sin(\dfrac{\pi}{\Phi}w)}{k_o \sin(w)}$.

The second and last case is $\Phi/2 < \varphi_o < \Phi$. For this case, we have the incident and face a reflected waves that generate a non-standard part in minus functions; therefore, we need to consider the non-standard part of $\bar{X}^{n.s.}_{(i)-}(\bar{\eta})$ i = 1,2,3,4

$$\bar{X}_{(i)-}(\bar{\eta}) = \bar{X}^{s}_{(i)-}(\bar{\eta}) + \frac{\bar{U}_{oi}}{\bar{\eta} - \bar{\eta}_o}$$

whose definitions in the η plane are given in section 6.3

$$\begin{aligned} X_{1+}(-m) &= -\frac{n_{o-}}{\xi_-}\left[V_{a+}(-m\ ) + V_{b+}(-m\ )\right], \\ X_{2+}(-m) &= \frac{\omega\mu_o}{\xi_-^2}\left[I_{a+}(-m\ ) - I_{b+}(-m\ )\right] \end{aligned} \tag{6.B.15}$$

$$\begin{aligned} X_{3+}(-m) &= \frac{n_{o-}}{\sqrt{k_o}}\left[V_{a+}(-m\ ) - V_{b+}(-m\ )\right], \\ X_{4+}(-m) &= -\frac{\omega\mu_o}{\sqrt{k_o}\,\xi_-}\left[I_{a+}(-m\ ) + I_{b+}(-m\ )\right] \end{aligned} \tag{6.B.16}$$

Considering face b $\varphi = -\Phi$ in the shadow region, (6.B.15)–(6.B.16) contain non-standard components only of face a. We recall that

$$\bar{U}_{oi} = Res[\bar{X}_{(i)-}(\bar{\eta})\Big|_{\bar{\eta}_o} = U_{oi}\left.\frac{dm}{d\eta}\right|_{m_o} \tag{6.B.17}$$

with $U_{oi} = Res[X_{i-}(m)]\Big|_{m_o}$ and $m_o^{(+\Phi)} = k_o \cos(\Phi - \varphi_o)$ (incident wave pole in the m-plane for the spectra along $\varphi = +\Phi$). See generalization for the spectra on $\varphi = -\Phi$, as discussed previously.

Using the m-Laplace transform on the incident and face a reflected wave for the E_z components, we have

$$V_{z+}^{g}(-m,\Phi)=\int_0^{\infty}\left(E_o+E_{or}^{(a)}\right)e^{jk_o\rho\cos(\Phi-\varphi_o)}e^{-jm\rho}d\rho=\frac{-jE_o\left(1+\Gamma_a^{\perp}\right)}{m-m_o^{(+\Phi)}} \qquad (6.B.18)$$

with the reflection coefficient defined as

$$E_{or}^{(a)}=\Gamma_a^{\perp}E_o=\frac{Z_1\sin(\Phi-\varphi_o)-Z_o\sin(\Phi-\varphi_{ta})}{Z_1\sin(\Phi-\varphi_o)+Z_o\sin(\Phi-\varphi_{ta})}E_o \qquad (6.B.19)$$

where $Z_1=Z_o/\sqrt{\varepsilon_r}$, $\varphi_{ta}=\Phi+g(\Phi-\varphi_o)$ and $w_1=g(w)=-\arccos\left(\frac{\cos w}{\sqrt{\varepsilon_r}}\right)$ (Snell's law).

Using the definitions of the involved functions in the w plane (6.B.9), we obtain for (6.B.17) in detail

$$\bar{U}_{o1}=Res[\bar{X}_{(1)-}(\bar{\eta})\Big|_{\bar{\eta}_o}=U_{o1}\frac{d\bar{\eta}}{dm}\Big|_{m_o}$$

$$=-\frac{k_o\sin(w+\Phi)}{\sqrt{k_o}\cos(\frac{\pi w}{2\Phi})}\frac{1}{\sqrt{k_o}\sin(\frac{\pi w}{2\Phi})}(-jE_o\left(1+\Gamma_a^{\perp}\right))\frac{k_o\frac{\pi}{\Phi}\sin(\frac{\pi}{\Phi}w)}{-k_o\sin(w+\Phi)}\Bigg|_{m_o,w_o,\bar{\eta}_o}$$

$$=-j2\frac{\pi}{\Phi}E_o\left(1+\Gamma_a^{\perp}\right)$$

$$\bar{U}_{o3}=Res[\bar{X}_{(3)-}(\bar{\eta})\Big|_{\bar{\eta}_o}=U_{o3}\frac{d\bar{\eta}}{dm}\Big|_{m_o}$$

$$=\frac{k_o\sin(w+\Phi)}{\sqrt{k_o}\cos(\frac{\pi w}{2\Phi})}(-jE_o\left(1+\Gamma_a^{\perp}\right))\frac{k_o\frac{\pi}{\Phi}\sin(\frac{\pi}{\Phi}w)}{-k_o\sin(w+\Phi)}\Bigg|_{m_o,w_o,\bar{\eta}_o}$$

$$=-j2\frac{\pi}{\Phi}E_o\left(1+\Gamma_a^{\perp}\right)\sqrt{k_o}\sin(\frac{\pi\varphi_o}{2\Phi})$$

(6.B.20)

Using the m-Laplace transform on the incident and face a reflected waves for the H_ρ components (5.2.2) (valid for total field in general) and (6.B.2)

$$H_\rho^i(\rho,\varphi) = \frac{E_o}{Z_o}\sin(\varphi-\varphi_o)e^{jk_o\rho\cos(\varphi-\varphi_o)} \tag{6.B.21}$$

$$H_\rho^{ra}(\rho,\varphi) = \Gamma_a^\perp\frac{E_o}{Z_o}\sin(\varphi+\varphi_o-2\Phi_o)e^{jk_o\rho\cos(\varphi+\varphi_o-2\Phi_o)} \tag{6.B.22}$$

we have

$$\begin{aligned} I_{\rho+}^g(-m,\Phi) &= \frac{\sin(\Phi-\varphi_o)E_o\left(1-\Gamma_a^\perp\right)}{Z_o}\int_0^\infty e^{jk_o\rho\cos(\Phi-\varphi_o)}e^{-jm\rho}d\rho \\ &= \frac{-jE_o\left(1-\Gamma_a^\perp\right)}{m-m_o^{(+\Phi)}}\frac{\sin(\Phi-\varphi_o)}{Z_o} \end{aligned} \tag{6.B.23}$$

and thus

$$\begin{aligned} \bar{U}_{o2} &= Res[\bar{X}_{(2)-}(\bar{\eta})\Big|_{\bar{\eta}_o} \\ &= U_{o2}\frac{d\bar{\eta}}{dm}\bigg|_{m_o} = \left.\frac{\omega\mu_o}{k_o\sin^2(\frac{\pi w}{2\Phi})}\left(-jE_o\left(1-\Gamma_a^\perp\right)\frac{\sin(\Phi-\varphi_o)}{Z_o}\right)\frac{k_o\frac{\pi}{\Phi}\sin(\frac{\pi}{\Phi}w)}{-k_o\sin(w+\Phi)}\right|_{m_o,w_o,\bar{\eta}_o} \\ &= -2jE_o\left(1-\Gamma_a^\perp\right)\frac{\pi}{\Phi}\cot(\frac{\pi}{2\Phi}\varphi_o) \\ \bar{U}_{o4} &= Res[\bar{X}_{(4)-}(\bar{\eta})\,|_{\bar{\eta}_o} \\ &= U_{o4}\frac{d\bar{\eta}}{dm}\bigg|_{m_o} = \left.-\frac{\omega\mu_o}{k_o\sin(\frac{\pi w}{2\Phi})}\left(-jE_o\left(1-\Gamma_a^\perp\right)\frac{\sin(\Phi-\varphi_o)}{Z_o}\right)\frac{k_o\frac{\pi}{\Phi}\sin(\frac{\pi}{\Phi}w)}{-k_o\sin(w+\Phi)}\right|_{m_o,w_o,\bar{\eta}_o} \\ &= -2jE_o\left(1-\Gamma_a^\perp\right)\frac{\pi}{\Phi}\cos(\frac{\pi}{2\Phi}\varphi_o) \end{aligned} \tag{6.B.24}$$

With respect to Daniele and Lombardi (2011), we obtain $\bar{T}_{o3}$ (6.B.14), $\bar{U}_{o3}$ (6.B.20) and $\bar{U}_{o4}$ (6.B.24) with different signs since the definitions of the plus and minus unknowns with i = 3,4 are changed in sign, but these redefinitions yield the same GWHEs (6.3.2)–(6.3.5) and (6.A.1)–(6.A.4) for coherence.

Now the explicit form of (6.B.8) becomes

$$\bar{n}_i(\bar{\eta}) = \begin{cases} \dfrac{\bar{T}_{oi}}{\bar{\eta}-\bar{\eta}_o} & 0<\varphi_o<\Phi/2 \\ \dfrac{\bar{U}_{oi}}{\bar{\eta}-\bar{\eta}_o} + \dfrac{k_o+\bar{\eta}}{\sqrt{k_o^2-\bar{\eta}^2}}\dfrac{\bar{U}_{o(i+1)}}{\bar{\eta}-\bar{\eta}_o} & \Phi/2<\varphi_o<\Phi \end{cases} \tag{6.B.25}$$

with i = 1,3 and $\bar{T}_{oi}$, $\bar{U}_{oi}$ and $\bar{U}_{o(i+1)}$ defined in (6.B.14), (6.B.20) and (6.B.24).

References

Antipov, Y.A. and Silvestrov, V.V. (2004a). Vector functional-difference equation in electromagnetic scattering. *IMA Journal of Applied Mathematics*, 69(1), 27–69.

Antipov, Y.A. and Silvestrov, V.V. (2004b). Second-order functional-difference equations. II: Scattering from a right-angled conductive wedge for E-polarization. *Quarterly Journal of Mechanics and Applied Mathematics*, 57(2), 267–313.

Antipov, Y.A. and Silvestrov, V.V. (2006). Electromagnetic scattering from an anisotropic half-plane at oblique incidence: The exact solution. *Quarterly Journal of Mechanics and Applied Mathematics*, 59, 211–251.

Bernard, J.M.L. (1997). Méthode analytique et transformées fonctionnelles pour la diffraction d'ondes par une singularité conique: Equation intégrale de noyau non oscillant pour le cas d'impedance constant. Technical Report CEA-R-5764, CEA, Saclay, France.

Bernard, J.M.L. (1998). Diffraction at skew incidence by an anisotropic impedance wedge in electromagnetic theory: A new class of canonical cases. *Journal of Physics A: Mathematical and General*, 31(2), 595–613.

Bernard, J.M.L. (2014). *Advanced Theory of the Diffraction by a Semi Infinite Impedance Cone*. Alpha Science International Ltd, Oxford.

Bucci, O. and Franceschetti, G. (1976). Electromagnetic scattering by a half plane with two face impedances. *Radio Science*, 11, 49–59.

Budaev, B. (1995). *Diffraction by Wedges*. Longman Scientific & Technical, Burnt Mill, Harlow.

Croisille, J.P. and Lebeau, G. (1999). *Diffraction by an Immersed Elastic Wedge*. Springer-Verlag, Berlin, Heidelberg.

Daniele, V.G. (1984a). On the solution of two coupled Wiener–Hopf equations. *SIAM Journal of Applied Mathematics*, 44, 667–680.

Daniele, V.G. (1984b). On the solution of vector Wiener–Hopf equations occurring in scattering problems. *Radio Science*, 19, 1173–1178.

Daniele, V.G. (2003a). Rotating waves in the Laplace domain for angular regions. *Electromagnetics*, 23, 223–236.

Daniele, V.G. (2003b). The Wiener–Hopf technique for impenetrable wedges having arbitrary aperture angle. *SIAM Journal of Applied Mathematics*, 63, 1442–1460.

Daniele, V.G. (2004). The Wiener–Hopf technique for wedge problems. Internal Report ELT-2004-2 [Online]. Available: http://personal.det.polito.it/vito.daniele.

Daniele, V.G. (2009). The Wiener–Hopf solution of the penetrable wedge problem. Internal Report ELT-2009 [Online]. Available: http://personal.delen.polito.it/vito.daniele/.

Daniele, V.G. (2010). The Wiener–Hopf formulation of the dielectric wedge problem. Part I. *Electromagnetics*, 30(8), 625–643.

Daniele, V.G. (2011a). The Wiener–Hopf formulation of the dielectric wedge problem. Part II. *Electromagnetics*, 31(1), 1–17.

Daniele, V.G. (2011b). The Wiener–Hopf formulation of the dielectric wedge problem. Part III. *Electromagnetics*, 31(8), 550–570.

Daniele, V.G. and Lombardi, G. (2005). The Wiener–Hopf technique for impenetrable wedge problems. *International Conference on Days on Diffraction*, Saint Petersburg, Russia.

Daniele, V.G. and Lombardi, G. (2006). Wiener–Hopf solution for impenetrable wedges at skew incidence. *IEEE Transactions on Antennas and Propagation*, 54(9), 2472–2485.

Daniele, V.G. and Lombardi, G. (2011). The Wiener–Hopf solution of the isotropic penetrable wedge problem: Diffraction and total field. *IEEE Transactions on Antennas and Propagation*, 59(10), 3797–3818.

Daniele, V.G. and Lombardi, G. (2016). Arbitrarily oriented perfectly conducting wedge over a dielectric half-space: Diffraction and total far field. *IEEE Transactions on Antennas and Propagation*, 64(4), 1416–1433.

Daniele, V.G. and Uslenghi, P.L.E. (2007). Isorefractive wedge. *Radio Science*, 42, 1–9.

Daniele, V.G. and Zich, R.S. (2014). *The Wiener–Hopf Method in Electromagnetics*. Scitech Publishing, Edison, New Jersey.

Daniele, V.G., Lombardi, G., and Zich, R.S. (2017a). Network representations of angular regions for electromagnetic scattering. *PLoS One*, 12(8), e0182763.

Daniele, V.G., Lombardi, G., and Zich, R.S. (2017b). The electromagnetic field for a PEC wedge over a grounded dielectric slab: 1. Formulation and validation. *Radio Science*, 52(12), 1472–1491.

Daniele, V.G., Lombardi, G., and Zich, R.S. (2017c). The electromagnetic field for a PEC wedge over a grounded dielectric slab: 2. Diffraction, modal field, surface waves, and leaky waves. *Radio Science*, 52(12), 1492–15098.

Daniele, V.G., Lombardi, G., and Zich, R.S. (2018). The double PEC wedge problem: Diffraction and total far field. *IEEE Transactions on Antennas and Propagation*, 66(12), 6470–6673.

Daniele, V.G., Lombardi, G., and Zich, R.S. (2019). Radiation and scattering of an arbitrarily flanged dielectric-loaded waveguide. *IEEE Transactions on Antennas and Propagation*, 67(12), 7569–7584.

Daniele, V.G., Lombardi, G., and Uslenghi, P.L.E. (2020). Scattering by a double negative (DNG) wedge. Internal Report 1, Politecnico di Torino.

Gakhov, F.D. (1966). *Boundary Value Problems*. Dover Publications, Inc., New York.

Gautesen, A.K. (2001). Scattering of a Rayleigh wave by an elastic wedge whose angle is greater than 180 degrees. *ASME Journal of Applied Mechanics*, 68, 476–479.

Gautesen, A.K. and Fradkin, L.J. (2010). Diffraction by a two-dimensional traction-free elastic wedge. *SIAM Journal of Applied Mathematics*, 70, 3065–3085.

Hurd, R.A. and Lüneburg, E. (1985). Diffraction by an anisotropic impedance half plane. *Canadian Journal of Physics*, 63(9), 1135–1140.

Kantorovich, L.V. and Krylov, V.I. (1964). *Approximate Methods of Higher Analysis*. Noordhoff, Groningen.

Kim, S.Y., Ra, J.W., and Shin, S.Y. (1991a). Diffraction by an arbitrary – angled dielectric wedge: I – Physical optics approximation. *IEEE Transactions on Antennas and Propagation*, 39(9), 1272–1281.

Kim, S.Y., Ra, J.W., and Shin, S.Y. (1991b). Diffraction by an arbitrary – angled dielectric wedge: II – Correction to physical optics solution. *IEEE Transactions on Antennas and Propagation*, 39(9), 1282–1292.

Knockaert, L., Olyslager, F., and De Zutter, D. (1997). The diaphaneous wedge. *IEEE Transactions on Antennas and Propagation*, AP-45, 1374–1381.

Kouyoumjian, R.G. and Pathak, P.H. (1974). A uniform geometrical theory of diffraction for an edge in a perfectly conducting surface. *Proceedings of the IEEE*, 62(11), 1448–1461.

Lombardi, G. (2012). Skew incidence on concave wedge with anisotropic surface impedance. *IEEE Antennas and Wireless Propagation Letters*, 11, 1141–1145.

Lüneburg, E. and Serbest, A.H. (2000). Diffraction of an obliquely incident plane wave by a two-face impedance half plane: Wiener–Hopf approach. *Radio Science*, 35, 1361–1374.

Lyalinov, M.A. and Zhu, N.Y. (1999a). Diffraction of a normally incident plane wave at a wedge with identical tensor impedance surfaces. *IEEE Transactions on Antennas and Propagation*, 47, 914–917.

Lyalinov, M.A. and Zhu, N.Y. (1999b). Diffraction of a skewly incident plane wave by an anisotropic impedance wedge – a class of exactly solvable cases. *Wave Motion*, 30(3), 275–288.

Lyalinov, M.A. and Zhu, N.Y. (2003a). A solution procedure for second order difference equations and its application to electromagnetic wave diffraction in a wedge-shaped region. *Proceedings of the Royal Society A*, 459(2040), 3159–3180.

Lyalinov, M.A. and Zhu, N.Y. (2003b). Exact solution to diffraction problem by wedges with a class of anisotropic impedance faces: Oblique incidence of a plane electromagnetic wave. *IEEE Transactions on Antennas and Propagation*, 51, 1216–1220.

Lyalinov, M.A. and Zhu, N.Y. (2013). *Scattering of Waves by Wedges and Cones with Impedance Boundary Conditions*. Scitech Publishing, Edison, New Jersey.

Malyuzhinets, G.D. (1958). Excitation, reflection and emission of surface waves from a wedge with given face impedances. *Soviet Physics, Doklady*, 3, 752–755.

Manara, G. and Nepa, P. (2000). Electromagnetic diffraction of an obliquely incident plane wave by a right-angled anisotropic impedance wedge with a perfectly conducting face. *IEEE Antennas and Propagation*, 48(4), 547–555.

Manara, G., Nepa, P., and Pelosi, G. (2000). High frequency EM scattering by edges in artificially hard and soft surfaces illuminated at oblique incidence. *IEEE Transactions on Antennas and Propagation*, 48(5), 790–800.

Manara, G., Nepa, P., Pelosi, G., and Vallecchi, A. (2004). Skew incidence diffraction by an anisotropic impedance half plane with a PEC face and arbitrarily oriented anisotropy axes. *IEEE Transactions on Antennas and Propagation*, 52(2), 487–496.

Nepa, P., Manara G., and Armogida, A. (2000). Electromagnetic scattering by anisotropic impedance half and full planes illuminated at oblique incidence. *IEEE Transactions on Antennas and Propagation*, 49(5), 790–800.

Norris, A.N. and Osipov, A.V. (1999). Far field analysis of the Malyuzhinets solution for plane and surface waves diffraction by an impedance wedge. *Wave Motion*, 30, 69–89.

Peters, A.S. (1952). Water waves over sloping beaches and the solution of a mixed boundary value problem for scalar Helmholtz equation in a sector. *Communications on Pure and Applied Mathematics*, 5, 97–108.

Rawlins, A.D. (1977). Diffraction by a dielectric wedge. *Journal of the Institute of Mathematics and its Applications*, 19, 261–279.

Rawlins, A.D. (1999). Diffraction by, or diffusion into, a penetrable wedge. *Proceedings of the Royal Society of London. Series A: Mathematical, Physical and Engineering Sciences*, 455, 2655–2686.

Rubinowicz, A. (1966). *Die Beugungswelle in der Kirchhoffschen Theorie der Beugung*. Springer-Verlag, Berlin.

Salem, M.A., Kamel, A., and Osipov, A.V. (2006). Electromagnetic fields in the presence of an infinite dielectric wedge. *Proceedings of the Royal Society of London. Series A: Mathematical, Physical and Engineering Sciences*, 462, 2503–2522.

Scharstein, R.W. (1993). Mellin transform solution for the static line-source excitation of a dielectric wedge. *IEEE Transactions on Antennas and Propagation*, AP-41, 1675–1679.

Sendag, R. and Serbest, H. (2001). Scattering at the junction formed by a PEC half-plane and a half-plane with anisotropic conductivity. *Electromagnetics*, 21, 415–434.

Senior, T.B.A. (1959). Diffraction by an imperfectly conducting wedge. *Communications on Pure and Applied Mathematics*, 12, 337–372.

Senior, T.B.A. (1978). Some problems involving imperfect half planes. In *Electromagnetic Scattering*, Uslenghi, P.L.E. (ed.). Academic Press, New York.

Senior, T.B.A. and Legault, S.R. (1998). Diffraction by an anisotropic impedance half-plane at skew incidence. *Electromagnetics*, 18, 207–225.

Senior, T.B.A. and Legault, S.R. (2000). Second-order difference equations in diffraction theory. *Radio Science*, 35(3), 683–690.

Senior, T.B.A. and Osipov, A.V. (2007). Plane wave solutions for right-angled interior impedance wedges. *Radio Science*, 42(RS6S04), 1–8.

Senior, T.B.A. and Volakis, J.L. (1995). *Approximate Boundary Conditions in Electromagnetics*. Institution of Electrical Engineers, London.

Shanin, A.V. (1996). On wave excitation in a wedge-shaped region. *Acoustical Physics*, 42(5), 696–701.

Sommerfeld, A. (1896). Mathematische theorie der diffraction. *Mathematische Annalen*, 47, 317–341.

Vaccaro, V. (1981). Electromagnetic diffraction from a right-angled wedge with soft conditions on one face. *Optica Acta*, 28, 293–311.

Williams, W.E. (1959). Diffraction of an e-polarized plane wave by an imperfectly conducting wedge. *Proceedings of the Royal Society of London. Series A*, 252(1270), 376–393.

Zhu, N.Y. and Landstorfer, F.M. (1997). Numerical study of diffraction and slope-diffraction at anisotropic impedance wedges by the method of parabolic equation: Space waves. *IEEE Transactions on Antennas and Propagation*, 45(5), 822–828.

Index

A, B, C

analytic
- continuation, 100, 101, 147, 154, 156, 159, 161, 178, 181
- extension, 99, 101, 106, 115, 117

angle
- azimuthal, 110, 140, 176
- observation, 105, 107, 122–125
- zenithal, 110

angular complex plane, 88
anisotropic face, 124
approximate representations, 94, 97, 100, 101
axial spectra, 104, 105, 121, 139, 155, 157, 159, 162–164, 173, 175, 178, 182, 183, 189, 190
boundaries
- conditions, 4–6, 12, 16, 39, 43, 45, 58, 60, 62, 67, 73, 137–139, 147, 148
- shadow, 105, 107, 108, 130, 178

boundedness of the norm, 133
branch, 91, 92, 94–96, 106, 109, 142, 156, 157, 165
Bromwich, 104, 106
Cauchy formula, 138, 139, 150
co-polar, 124, 125, 127, 130, 131
compact operator, 159
compatibility conditions, 130, 131
complex plane $\overline{w}$ and $\overline{w}_d$, 144, 188
complex scattering problem, 129, 133
contour warping, 94
cross-polar, 128

D, E, F

Daniele–Khrapkov, 20, 21, 31, 37, 51, 54, 71, 75, 76
decomposition, 33, 35, 56, 57, 60, 62, 66, 68, 76, 79, 80, 82
- logarithmic, 4, 24, 41, 43, 81

diffraction, 105, 108, 110, 121–126, 131
eigenvalues, 3, 25, 27, 33, 38
exact solution, 1, 3, 4, 11, 18, 24, 26, 27, 31, 36, 37, 39, 40, 51, 62, 68, 72, 87, 98, 132
Ez-polarization, 140–142, 154, 176, 177
factorization, 1, 2, 4, 9–11, 14, 15, 18, 20–22, 24, 26, 27, 30, 32, 35–38, 40, 41, 43, 47–51, 54–57, 59, 60, 62, 64–72, 77–81, 83
- Fredholm, 10, 51, 58, 62, 86, 92, 95, 135, 136, 138, 144, 152, 159, 164, 165, 176, 186, 187
- weak, 27, 31

far field, 101, 103, 104, 111, 113, 117, 126–128, 130, 160, 164, 177, 178

total, 126
pattern, 177
field evaluation, 162, 164
Fredholm integral equation of second kind (FIE), 88, 92, 94, 95, 97, 110, 112, 119, 129, 133, 135, 137, 144, 146, 150, 152, 154, 159, 165, 176, 188
function
$g(w)$ and $g_1(w_1)$, 156, 157
minus, 136–138, 142, 144, 145, 189–191, 195
multivalued, 91, 95, 100, 156
plus, 139, 142, 144, 145, 157, 182, 187, 189, 193
sectional analytic, 154

G, H, I

Geometrical Optics (GO), 92, 94, 101, 105–108, 117–120, 122, 127–130, 146, 163, 177, 178, 182, 183, 185, 189–192
regions, 178
GTD, 105, 107, 108, 110, 121–126, 131, 164, 181–184
half-plane, 4, 8, 9, 11, 13, 15, 18, 24, 31, 32, 35–41
penetrable, 10, 39
homogeneous medium, 87, 88
impedances, 87
anisotropic, 24, 28, 41, 42, 94, 124
isotropic, 18, 52
incidence
normal, 9, 10, 31, 32, 38, 41, 51, 52, 64, 67, 68, 70, 71, 138, 140, 141, 165, 173, 176, 177, 186
skew, 5, 9–11, 18, 26–28, 36, 37, 41, 42, 51, 58, 60, 67, 70, 72, 87–89, 94, 103, 121, 124–126, 128, 140, 165, 176, 186
interface, 138, 142, 183

J, K, L

Jordan representation, 3
kernel, 90, 91, 95, 97, 99–103, 105, 109, 111–113, 116, 121, 126, 132, 133, 135, 140, 147, 152, 154, 155, 159–162
Kontorovich–Lebedev (KL), 1, 36
Kouyoumjian–Pathak (KP), 107, 183
Laplace
domain, 88
spectra, 93
transforms, 135, 139–141, 159, 173, 190, 192
inverse, 104, 108, 116
m-, 191, 196, 197
Leontovich, 6
linear system, 98–100, 113, 125
Liouville theorem, 22

M, N, O

machine precision, 112–115
Malyuzhinets
function, 57, 86
problem, 10, 52, 67, 80, 86
mapping, 91, 95, 97, 100, 102, 136, 138, 139, 142, 156, 183, 185, 186
Maxwell equations, 6, 11, 43
Meixner's condition, 132, 133, 152, 160
multiple reflections, 93, 129
multiple wave numbers, 135
non-standard, 22, 24, 57, 92–94, 128, 129, 146, 187, 189–193, 195
normal form, 89
normalized values, 159
observation point, 96
offending exponential behavior, 24

P, Q, R

PEC, 28, 43, 46, 58, 60, 102, 118–121, 123
penetrable, 10, 39
permeability, 137, 140
permittivity, 137, 140
piecewise constant collocation method, 97, 99

plane junctions, 4, 9–11, 37
plus and minus, 91, 92, 129
PMC, 58
poles, 92–94, 104–106, 108, 109, 113, 115, 116, 119, 130
  fictitious, 113, 114
polynomial matrix, 4, 9, 11, 15, 18, 24, 27, 30, 32, 41, 48, 50, 64
propagation constant, 137, 138, 145, 159, 176, 189
proper/improper sheets, 91–96, 99, 101, 102, 116, 129
quadratically integrable, 133
reflection coefficients, 94, 118, 146, 196
regularity segment, 178–180, 185
relative error, 111, 113–116, 122–124, 126, 160, 176, 179, 182, 184
residue theorem, 25, 84

S, T, U

saddle point, 104, 105, 108
sample-and-hold quadrature method, 97, 161
scaling, 110, 113–116, 122–124, 126
scattering, 87–89, 95, 102, 103, 110, 132
SDP, 104–106, 108, 109, 121, 182
singularity lines, 99–103, 116, 147, 154, 155
Snell's law, 92, 108
Sommerfeld
  function, 104, 107, 109, 122
  integral, 107, 108
Sommerfeld–Malyuzhinets (SM), 1, 3, 4, 10, 11, 18, 26, 28, 36, 39–41, 43, 51, 52, 58, 60, 68, 70, 72, 76, 136
source, 146, 171, 187–189
  term, 146, 188, 189
strip, 102, 103, 154–156, 173, 183
  initial validity, 99, 101–103, 106
  regularity, 161, 179, 180
technical computing software, 88
uniformization, 108, 121
UTD, 107, 118–120, 127, 128, 130, 163, 177, 178, 183, 185

V, W

validity, 103
Watson's lemma, 132
wave
  plane, 87–89, 92, 103, 110, 119, 132, 140, 165, 176, 177, 186, 187, 189
  surface (SW wave), 105, 108–110, 119–121
  transmitted, 177, 178, 182, 190, 191
wedge, 1, 87
  concave, 128–131
  dielectric, 136–138, 140–142, 154, 159, 160, 162–165, 173, 176, 177, 183, 186
  penetrable, 135
  right, 4, 10, 11
  right-angled, 40–42
Wiener-Hopf (WH)
  equation, 1, 3, 5, 10, 11, 21, 29, 31, 37–40, 43, 46, 51, 65, 67, 69, 74–76
    generalized (GWHE), 5, 8, 51–54, 58, 62, 65, 77, 88, 92, 95, 97, 101, 135, 136, 139, 140, 142, 155, 159, 162, 186, 198
  technique, 1, 3, 9, 18, 43, 51, 52, 60

Summary of Volume 1

Preface

Introduction

Chapter 1. Introduction to the Wiener-Hopf Method

1.1. A brief history of the Wiener-Hopf method
1.2. Fundamental definitions and assumptions to develop the Wiener-Hopf technique in the spectral domain
1.3. WH equations from the physical domain to the spectral domain
1.4. WH equations in wave scattering problems
1.5. Classical solution of WH equations
1.6. The decomposition (Cauchy) equations
1.7. General formula for factorization of scalar kernels
1.8. Some explicit factorization of matrix kernels useful in wedge scattering
1.9. The Fredholm factorization technique: a general technique to solve WH equations
1.10. Spectral properties of the unknowns
1.11. Semi-analytical solution of the Fredholm integral equation
1.12. Analytic continuation outside the integration line in the η-plane
1.13. The complex plane w
1.14. The WH unknowns in the w plane
1.15. The Fredholm factorization technique in the w plane
1.16. Analytic continuation in the w plane
1.17. The Fredholm factorization technique to factorize the kernel
1.18. Some examples of the Fredholm factorization method
Appendix 1.A: effect of using j and i as imaginary units in Fourier transforms
Appendix 1.B: compactness of the matrix kernel $\frac{G^{-1}(\alpha)G(u)-\mathbf{1}}{u-\eta}$

Chapter 2. A Basic Example: Scattering by a Half-plane

2.1. The fundamental problem of diffraction in wave motion
2.2. Unified theory of transverse equations in arbitrary stratified regions
2.3. WH equation for electromagnetic diffraction by a perfectly electrically conducting (PEC) half-plane illuminated by an Ez-polarized plane wave at normal incidence
2.4. Non-standard contributions of WH unknowns in the PEC half-plane illuminated by an Ez-polarized plane wave at normal incidence
2.5. Solution of the WH equation of the PEC half-plane using classical closed-form factorization
2.6. Solution of the WH equation of the PEC half-plane using the Fredholm factorization method
2.7. Field estimation
2.8. Numerical validation of the Fredholm factorization method
2.9. Generality of the wave motion in a homogeneous isotropic elastic solid
2.10. Plane waves in a homogeneous elastic solid and simplifications
2.11. Diffraction by a half-infinite crack in a homogeneous elastic solid planar problem
2.12. Diffraction by a half-infinite crack in a homogeneous isotropic two-dimensional elastic solid problem

Chapter 3. The Wiener-Hopf Theory for Angular Region Problems

3.1. A brief history of the classical methods for studying angular regions
3.2. Introduction to the generalized Wiener-Hopf technique
3.3. WH functional equations in the angular region filled by a homogeneous isotropic medium in Electromagnetics
3.4. Reduction of the generalized functional equations of an angular region to functional equations defined in the same complex plane
3.5. Generalized Wiener-Hopf equations for the impenetrable wedge scattering problem
3.6. Solution of generalized Wiener-Hopf equations for the impenetrable wedge scattering problem
3.7. Non-standard parts of the plus and minus functions in GWHEs for the impenetrable wedge scattering problem
3.8. Closed-form solution of the PEC wedge scattering problem at normal incidence
3.9. Alternative solution of the PEC wedge scattering problem at normal incidence via difference equations
3.10. Generalized WH functional equations for angular regions in the w plane
3.11. Rotating wave method
3.12. Properties of rotating waves

3.13. Spectral field component in w for an arbitrary direction
φ using rotating waves
3.14. Rotating waves in areas different from electromagnetism
3.15. Closed-form solution of the diffraction of an elastic SH
wave by wedge with classical factorization
3.16. Rotating waves with the MF transform for wedge problems
3.17. Alternative solution of PEC wedge scattering problems via
difference equations and the MF transform in terms of rotating waves
Appendix 3.A: the Malyuzhinets–Fourier (MF) transform

References

Other titles from

iSTE

in

Waves

2020

SAKHO Ibrahima
Introduction to Quantum Mechanics 2: Wave-Corpuscle, Quantization & Schrödinger's Equation

2019

BERTRAND Pierre, DEL SARTO Daniele, GHIZZO Alain
The Vlasov Equation 1: History and General Properties

DAHOO Pierre-Richard, LAKHLIFI Azzedine
Infrared Spectroscopy of Triatomics for Space Observation
(Infrared Spectroscopy Set – Volume 2)

RÉVEILLAC Jean-Michel
Electronic Music Machines: The New Musical Instruments

ROMERO-GARCIA Vicente, HLADKY-HENNION Anne-Christine
Fundamentals and Applications of Acoustic Metamaterials: From Seismic to Radio Frequency
(Metamaterials Applied to Waves Set – Volume 1)

SAKHO Ibrahima
Introduction to Quantum Mechanics 1: Thermal Radiation and Experimental Facts Regarding the Quantization of Matter

2018

SAKHO Ibrahima
Screening Constant by Unit Nuclear Charge Method: Description and Application to the Photoionization of Atomic Systems

2017

DAHOO Pierre-Richard, LAKHLIFI Azzedine
Infrared Spectroscopy of Diatomics for Space Observation (Infrared Spectroscopy Set – Volume 1)

PARET Dominique, HUON Jean-Paul
Secure Connected Objects

PARET Dominque, SIBONY Serge
Musical Techniques: Frequencies and Harmony

RÉVEILLAC Jean-Michel
Analog and Digital Sound Processing

STAEBLER Patrick
Human Exposure to Electromagnetic Fields

2016

ANSELMET Fabien, MATTEI Pierre-Olivier
Acoustics, Aeroacoustics and Vibrations

BAUDRAND Henri, TITAOUINE Mohammed, RAVEU Nathalie
The Wave Concept in Electromagnetism and Circuits: Theory and Applications

PARET Dominique
Antennas Designs for NFC Devices

PARET Dominique
Design Constraints for NFC Devices

WIART Joe
Radio-Frequency Human Exposure Assessment

2015

PICART Pascal
New Techniques in Digital Holography

2014

APPRIOU Alain
Uncertainty Theories and Multisensor Data Fusion

JARRY Pierre, BENEAT Jacques N.
RF and Microwave Electromagnetism

LAHEURTE Jean-Marc
UHF RFID Technologies for Identification and Traceability

SAVAUX Vincent, LOUËT Yves
MMSE-based Algorithm for Joint Signal Detection, Channel and Noise Variance Estimation for OFDM Systems

THOMAS Jean-Hugh, YAAKOUBI Nourdin
New Sensors and Processing Chain

TING Michael
Molecular Imaging in Nano MRI

VALIÈRE Jean-Christophe
Acoustic Particle Velocity Measurements using Laser: Principles, Signal Processing and Applications

VANBÉSIEN Olivier, CENTENO Emmanuel
Dispersion Engineering for Integrated Nanophotonics

2013

BENMAMMAR Badr, AMRAOUI Asma
Radio Resource Allocation and Dynamic Spectrum Access

BOURLIER Christophe, PINEL Nicolas, KUBICKÉ Gildas
Method of Moments for 2D Scattering Problems: Basic Concepts and Applications

GOURE Jean-Pierre
Optics in Instruments: Applications in Biology and Medicine

LAZAROV Andon, KOSTADINOV Todor Pavlov
Bistatic SAR/GISAR/FISAR Theory Algorithms and Program Implementation

LHEURETTE Eric
Metamaterials and Wave Control

PINEL Nicolas, BOURLIER Christophe
Electromagnetic Wave Scattering from Random Rough Surfaces: Asymptotic Models

SHINOHARA Naoki
Wireless Power Transfer via Radiowaves

TERRE Michel, PISCHELLA Mylène, VIVIER Emmanuelle
Wireless Telecommunication Systems

2012

LALAUZE René
Chemical Sensors and Biosensors

LE MENN Marc
Instrumentation and Metrology in Oceanography

LI Jun-chang, PICART Pascal
Digital Holography

2011

BECHERRAWY Tamer
Mechanical and Electromagnetic Vibrations and Waves

BESNIER Philippe, DÉMOULIN Bernard
Electromagnetic Reverberation Chambers

GOURE Jean-Pierre
Optics in Instruments

GROUS Ammar
Applied Metrology for Manufacturing Engineering

LE CHEVALIER François, LESSELIER Dominique, STARAJ Robert
Non-standard Antennas

2010

BEGAUD Xavier
Ultra Wide Band Antennas

MARAGE Jean-Paul, MORI Yvon
Sonar and Underwater Acoustics

2009

BOUDRIOUA Azzedine
Photonic Waveguides

BRUNEAU Michel, POTEL Catherine
Materials and Acoustics Handbook

DE FORNEL Frédérique, FAVENNEC Pierre-Noël
Measurements using Optic and RF Waves

FRENCH COLLEGE OF METROLOGY
Transverse Disciplines in Metrology

2008

FILIPPI Paul J.T.
Vibrations and Acoustic Radiation of Thin Structures

LALAUZE René
Physical Chemistry of Solid-Gas Interfaces

2007

KUNDU Tribikram
Advanced Ultrasonic Methods for Material and Structure Inspection

PLACKO Dominique
Fundamentals of Instrumentation and Measurement

RIPKA Pavel, TIPEK Alois
Modern Sensors Handbook

2006

BALAGEAS Daniel *et al.*
Structural Health Monitoring

BOUCHET Olivier *et al.*
Free-Space Optics

BRUNEAU Michel, SCELO Thomas
Fundamentals of Acoustics

FRENCH COLLEGE OF METROLOGY
Metrology in Industry

GUILLAUME Philippe
Music and Acoustics

GUYADER Jean-Louis
Vibration in Continuous Media

Printed and bound by CPI Group (UK) Ltd, Croydon, CR0 4YY

15/09/2026

14962360-0002